# Laplace Transforms

**VNR NEW MATHEMATICS LIBRARY**

under the general editorship of

J.V. Armitage

*Principal*
*College of St. Hild and St. Bede*
*Durham University*

N. Curle

*Professor of Applied Mathematics*
*University of St. Andrews*

The aim of this series is to provide a reliable modern coverage of those mainstream topics that form the core of mathematical instruction in universities and comparable institutions. Each book deals concisely with a well defined key area in pure or applied mathematics or statistics. Many of the volumes are intended not solely for students of mathematics, but also for engineering and science students whose training demands a firm grounding in mathematical methods.

**Titles in the series:**

1. Applied Differential Equations—N. Curle
3. Vector Methods—R.J. Cole
4. Boolean Algebra and its Uses—G. South
5. Basic Pure Mathematics II—A.S.T. Lue
6. Elementary Partial Differential Equations—R.J. Gribben
7. A First Course in Abstract Algebra—P.J. Higgins
8. Dynamics—W.E. Williams
9. Linear Algebra: an Introduction—A.O. Morris
10. Laplace Transforms and Applications—E.J. Watson

# Laplace Transforms

## and Applications

E.J. Watson
*Department of Mathematics*
*University of Manchester*

VAN NOSTRAND REINHOLD COMPANY
New York—Cincinnati—Toronto—London—Melbourne

**Published by Van Nostrand Reinhold Company Ltd.,
Molly Millars Lane, Wokingham, Berkshire, England**

*Published in 1981 by Van Nostrand Reinhold Company,
A Division of Litton Educational Publishing Inc.,
135 West 50th Street, New York, NY 10020, USA*

*Van Nostrand Reinhold Limited,
1410 Birchmount Road, Scarborough, Ontario, M1P 2E7,
Canada*

*Van Nostrand Reinhold Australia Pty. Limited,
17 Queen Street, Mitcham, Victoria 3132, Australia*

**Library of Congress Cataloging in Publication Data**

Watson, Eric John, 1924—
Laplace transforms and applications.

(VNR new mathematics library; 10)
Bibliography: P.
1. Laplace transformation. I. Title.
QA432. W37 515.7′23 80-12633
ISBN 0-442-30176-6
ISBN 0-442-30428-5 (pbk.)

*Photoset by Thomson Press (I) Limited New Delhi and
Printed in Great Britain by Henry Ling Ltd., Dorchester, Dorset.*

# Contents

# Introduction

The Laplace transformation is a powerful method for the solution of linear differential equations. Although Laplace had used integrals involving exponential functions for this purpose at the beginning of the 19th century, the method described here was developed about 100 years later in connection with Heaviside's operational calculus.

The aim of this work is to show how the Laplace transformation can be applied to a variety of problems. Laplace transforms have many useful properties, and I have attempted to give valid mathematical proofs of these. Many of the properties are quite easy to establish, but some are much harder to prove rigorously. Readers interested only in the applications may wish to pass over the more difficult proofs; others may prefer to leave them for a later reading. Passages which may well be omitted or postponed are marked by a line in the margin; these passages are intended primarily for reference purposes. Scientists and engineers should be aware that mathematical tools have limitations: mathematicians (who are not necessarily distinct persons from these scientists and engineers) should know why these limitations are there in order to give advice in cases of difficulty. This is not intended as a book on analysis, and those whose interest is mainly in the theory should consult works such as those of Doetsch (1974), Erdélyi (1962), Titchmarsh (1937) and Widder (1941) listed as references.

The material of this book has been arranged by subject into five chapters, but I do not suggest that these should be read straight through from beginning to end. Cross-references have been provided in order to encourage readers to go both forwards and backwards. A first course on Laplace transforms might consist of the elementary properties in Section 1.1 with their applications to differential equations with constant coefficients in Chapter 2, omitting parts marked by the margin line. Such a course would require a knowledge of the algebraic techniques used for partial fractions and simultaneous equations, as well as elementary calculus. For the more advanced applications, which require the use of the inversion theorem, the theory of functions of a complex variable is an essential tool.

The examples have been chosen to illustrate the uses of the Laplace transformation and the techniques needed in its application, rather than for any intrinsic interest. Additional examples for the reader to solve are provided at the end of each chapter. Most of these are problems to be solved by the methods described in the chapter, but some, marked with the marginal line, are meant to illustrate or develop the theory, or involve more difficult ideas. The Laplace transformation is not necessarily the best method of solution of the examples, whether worked out in the text or provided for the reader.

For ease of reference, the various definitions are labelled in the form D1.1.1, examples as E1.1.1, properties (or propositions) as P1.1.1 and equations as (1.1.1). Thus, the inversion theorem P3.2.7 is Property 7 of Section 2 of Chapter 3.

Lists of general properties of transforms and transforms of special functions are provided, together with references to the text. The second list should be used in conjunction with the first to obtain further transforms. Much more extensive lists are given by the Bateman Manuscript Project, edited by Erdélyi (1954). Further information on Laplace transforms and related subjects will be found in the references. I gladly acknowledge the help I have derived from these references, but I have often changed their arguments: errors in this book should not be imputed to other authors.

# 1 Laplace Transforms

## 1.1 Basic Properties

**D1.1.1** Let $u(t)$ be a function defined for $t > 0$. Its *Laplace transform* is defined to be

$$\mathscr{L}\{u(t)\} = \bar{u}(p) = \int_0^\infty u(t)\mathrm{e}^{-pt}\,\mathrm{d}t, \tag{1.1.1}$$

provided that the integral converges.

The Laplace transformation converts a function of $t$ into a function of the transform variable $p$. This may be thought of as real in the elementary applications of Laplace transforms, but, for the inversion theorem $p$ must be regarded as a complex variable. The notations $\mathscr{L}\{\ \}$ and $\bar{\ }$ will be used to denote transforms. Note that in a dimensional situation $p$ has the dimensions of $t^{-1}$ and $\bar{u}(p)$ has those of $tu(t)$. Some authors work with the function $p\bar{u}(p)$, which is Heaviside's *operational representation* of $u(t)$.

**E1.1.1**

$$\mathscr{L}\{1\} = \int_0^\infty \mathrm{e}^{-pt}\,\mathrm{d}t = p^{-1}.$$

**E1.1.2**

$$\mathscr{L}\{t\} = \int_0^\infty t\,\mathrm{e}^{-pt}\,\mathrm{d}t = p^{-2},$$

after integrating by parts.

**E1.1.3**

$$\begin{aligned}\mathscr{L}\{\cos\omega t\} &= \int_0^\infty \tfrac{1}{2}(\mathrm{e}^{\mathrm{i}\omega t} + \mathrm{e}^{-\mathrm{i}\omega t})\mathrm{e}^{-pt}\,\mathrm{d}t \\ &= \tfrac{1}{2}[(p-\mathrm{i}\omega)^{-1} + (p+\mathrm{i}\omega)^{-1}] = p/(p^2+\omega^2).\end{aligned}$$

**E1.1.5**

$$\mathscr{L}\{\sin \omega t\} = \int_0^\infty (1/2\mathrm{i})(\mathrm{e}^{\mathrm{i}\omega t} - \mathrm{e}^{-\mathrm{i}\omega t})\mathrm{e}^{-pt}\,\mathrm{d}t$$

$$= (1/2\mathrm{i})[(p - \mathrm{i}\omega)^{-1} - (p + \mathrm{i}\omega)^{-1}] = \omega/(p^2 + \omega^2).$$

In these examples $\alpha$ and $\omega$ are constants. The results of E1.1.1 and E1.1.2 are valid for $\mathscr{R}p > 0$, that of E1.1.3 for $\mathscr{R}p > \mathscr{R}\alpha$, and if $\omega$ is real then E1.1.4 and E1.1.5 hold for $\mathscr{R}p > 0$.

Although many simple functions, such as those in E1.1.1 to E1.1.5, have Laplace transforms, in other cases the integral (1.1.1) does not converge for any value of $p$. Examples include

(i) $u(t) = t^{-1}$,
(ii) $u(t) = \tan t$,
(iii) $u(t) = \exp(t^2)$,

for which the integral diverges as

(i) $t \to 0$,
(ii) $t \to \frac{1}{2}\pi, \frac{3}{2}\pi, \ldots$,
(iii) $t \to \infty$.

In order to give meaning to D1.1.1 and to develop the properties of transforms, we shall make certain assumptions about the functions involved.

**Assumption* 1 The function $u(t)$ is integrable in the sense of Riemann in any finite interval $0 \leqslant t \leqslant T$ with the possible exception of arbitrarily small neighbourhoods of a finite number of points $t_1, t_2, \ldots, t_N$, where $N$ may depend on $T$. Also

$$\int_0^T |u(t)|\,\mathrm{d}t$$

exists, if necessary as an improper integral, for every finite $T > 0$.

*Assumption* 2 There is a real constant $A$ such that

$$\int_0^T |u(t)|\mathrm{e}^{-At}\,\mathrm{d}t$$

converges as $T \to \infty$.

Assumption 1, which will always be made, may be expressed as '$u(t)$ is absolutely integrable in $0 \leqslant t \leqslant T$ for all $T > 0$', the

* Passages marked with a line in the margin may be omitted or postponed; see the Introduction.

phrase 'in the sense of Riemann' being understood. The most important case of an exceptional point is when $u(t)$ is unbounded as $t \to 0+$, so that $t_1 = 0$. Assumption 2 will also be made unless otherwise stated. The principal exceptions are in E1.5.30 to E1.5.32 and the uniqueness theorem P3.1.4.

It follows from Assumptions 1 and 2 that the Laplace transform $\bar{u}(p)$ exists and is uniformly convergent in $\mathscr{R}p \geqslant A$, since

$$\left| \int_T^\infty u(t) e^{-pt} \, dt \right| \leqslant \int_T^\infty |u(t)| \exp(-\mathscr{R}pt) \, dt \leqslant \int_T^\infty |u(t)| e^{-At} \, dt.$$

In several places $A$ will be used without formal definition in the sense of Assumption 2.

The functions $t^{-1}$ and $\tan t$ do not satisfy Assumption 1; $\exp(t^2)$ does not satisfy Assumption 2 for any $A$. Note that $\log|\sin t|$ satisfies Assumption 1 but $\log|\sin(t^{-1})|$ does not, although

$$\int_\delta^T \log|\sin(t^{-1})| e^{-pt} \, dt$$

has a limit when $T \to \infty$ and $\delta \to 0$, if $\mathscr{R}p > 0$.

In the following properties of Laplace transforms, it will be assumed that $\mathscr{R}p$ is always great enough for the various transforms to be absolutely convergent.

**P1.1.1** The Laplace transformation is a linear operation; if

$$\mathscr{L}\{u(t)\} = \bar{u}(p) \qquad \text{for } \mathscr{R}p \geqslant A$$

and

$$\mathscr{L}\{v(t)\} = \bar{v}(p) \qquad \text{for } \mathscr{R}p \geqslant B$$

then

$$\mathscr{L}\{\lambda u(t) + \mu v(t)\} = \lambda \bar{u}(p) + \mu \bar{v}(p)$$

for any constants $\lambda, \mu$, provided that $\mathscr{R}p \geqslant \max(A, B)$.

**P1.1.2** If $k > 0$ then

$$\mathscr{L}\{u(kt)\} = \int_0^\infty u(kt) e^{-pt} \, dt = \int_0^\infty u(s) e^{-ps/k} k^{-1} \, ds = k^{-1} \bar{u}(p/k).$$

Note that in general this is true only when $k$ is real and positive, and needs $\mathscr{R}p \geqslant kA$.

**P1.1.3** If $\alpha$ is a constant, then

$$\mathscr{L}\{e^{\alpha t}u(t)\} = \int_0^\infty u(t)e^{-(p-\alpha)t}\,dt = \bar{u}(p-\alpha).$$

**P1.1.4** If $u'(t) = du/dt$ exists for all $t > 0$ and has a Laplace transform, then

$$\mathscr{L}\{u'(t)\} = [u(t)e^{-pt}]_0^\infty - \int_0^\infty u(t)(-p)e^{-pt}\,dt = p\bar{u}(p) - u(0).$$

Here $u(0)$ means the limit of $u(t)$ as $t \to 0+$.

**P1.1.5** Repeated application of P1.1.4 gives

$$\mathscr{L}\{u''(t)\} = p^2\bar{u}(p) - pu(0) - u'(0),$$

and in general

$$\mathscr{L}\{u^{(n)}(t)\} = p^n\bar{u}(p) - p^{n-1}u(0) - p^{n-2}u'(0) - \ldots - u^{(n-1)}(0),$$

assuming that the Laplace transforms exist. The values of the derivatives $u'(0)$, etc., are the limiting values as $t \to 0+$.

**P1.1.6** If

$$v(t) = \int_0^t u(s)\,ds$$

then

$$\mathscr{L}\{v(t)\} = p^{-1}\bar{u}(p) \qquad \text{for } \mathscr{R}p > \max(A, 0).$$

This is the converse of P1.1.4, and is also proved by integrating by parts.

Since $v(t)$ is continuous, it satisfies Assumption 1. Also

$$|v(t)| \leqslant \int_0^t |u(s)|e^{-As}\cdot e^{As}\,ds$$

$$\leqslant \int_0^t |u(s)|e^{-As}\,ds(e^{At}+1) \leqslant K(e^{At}+1),$$

where $K$ is a constant. Thus if $\mathscr{R}p = x$

$$\int_0^T |v(t)|e^{-xt}\,dt \leqslant \int_0^T K(e^{At}+1)e^{-xt}\,dt,$$

so that Assumption 2 is satisfied if $x > \max(A, 0)$.

**E1.1.6** From E1.1.1 and P1.1.6, it follows by induction that

$$\mathscr{L}\{t^n\} = n!p^{-n-1} \qquad \text{for } n = 0, 1, 2, \dots \qquad \text{and } \mathscr{R}p > 0.$$

E1.1.2 is the case $n = 1$.

**D1.1.2** An *exponential-type function* (*e.t.f.*) is of the form

$$u(t) = \sum_{m=1}^{M} P_m(t) \exp(\alpha_m t), \tag{1.1.2}$$

where in each term $\alpha_m$ is a constant and

$$P_m(t) = \sum_{n=0}^{N_m} a_{mn} t^n \tag{1.1.3}$$

is a polynomial in $t$.

**P1.1.7** The sum and product of two e.t.f.s are e.t.f.s; the derivative and indefinite integral of an e.t.f. are e.t.f.s.

The first three results are obvious. For the last, note that

$$\int P(t)e^{\alpha t}\,dt = \sum_{n=0}^{N} (-1)^n \alpha^{-n-1} P^{(n)}(t) e^{\alpha t} \tag{1.1.4}$$

when $P(t)$ is a polynomial of degree $N$ in $t$.

Note that, as seen in E1.1.4 and E1.1.5, $\cos \omega t$ and $\sin \omega t$ are e.t.f.s, and hence so are their products with polynomials and exponentials. The properties P1.1.1 and P1.1.3, with E1.1.6, enable us to find the Laplace transform of any e.t.f.

**E1.1.7** If $u(t)$ is defined by equations (1.1.2) and (1.1.3), then

$$\bar{u}(p) = \sum_{m=1}^{M} \sum_{n=0}^{N_m} a_{mn} n! (p - \alpha_m)^{-n-1}. \tag{1.1.5}$$

Note that as $|p| \to \infty$

$$p\bar{u}(p) \to \sum_{m=1}^{M} a_{m0} = u(0).$$

**P1.1.8** If $\bar{u}(p)$ is a rational function of $p$ that tends to 0 as $|p| \to \infty$, so that

$$\bar{u}(p) = \frac{N(p)}{D(p)}$$

where $N(p)$ and $D(p)$ are polynomials in $p$ with the degree of $N(p)$ less than that of $D(p)$, then $\bar{u}(p)$ is the transform of an e.t.f.

For $\bar{u}(p)$ can be expressed in partial fractions in the form (1.1.5). Here the number $\alpha_m$ are the distinct roots (real or complex) of the equation $D(p)=0$, and $N_m+1$ is the multiplicity of the root $\alpha_m$.

**E1.1.8**

$$\frac{1}{p^3(p+1)^2}=\frac{1}{p^3}-\frac{2}{p^2}+\frac{3}{p}-\frac{1}{(p+1)^2}-\frac{3}{p+1}$$
$$=\mathscr{L}\{\tfrac{1}{2}t^2-2t+3-(t+3)e^{-t}\}.$$

**E1.1.9**

$$\begin{aligned}\bar{u}(p)&=\frac{p+3}{(p^2+2p+5)^2}\\&=\frac{p+3}{(p+1-2i)^2(p+1+2i)^2}\\&=-\frac{1+i}{8(p+1-2i)^2}-\frac{1-i}{8(p+1+2i)^2}\\&\quad-\frac{i}{16(p+1-2i)}+\frac{i}{16(p+1+2i)}\\&=\mathscr{L}\{u(t)\},\end{aligned}$$

where

$$\begin{aligned}u(t)&=[-\tfrac{1}{8}(1+i)t-\tfrac{1}{16}i]e^{(-1+2i)t}\\&\quad+[-\tfrac{1}{8}(1-i)t+\tfrac{1}{16}i]e^{(-1-2i)t}\\&=\tfrac{1}{4}e^{-t}[(t+\tfrac{1}{2})\sin 2t-t\cos 2t].\end{aligned}$$

Alternative derivations of these results are given as E1.4.2 and E1.4.7.

The properties of Laplace transforms given so far are enough to solve ordinary differential equations with constant coefficients, provided that initial conditions are given at $t=0$ and the forcing terms ('right-hand sides') of the equations are e.t.f.s. Equations of this type are treated in Sections 2.1 to 2.3.

**P1.1.9**

$$\mathscr{L}\{tu(t)\}=-\frac{d\bar{u}(p)}{dp}.$$

Formally, we have

$$-\frac{\mathrm{d}\bar{u}}{\mathrm{d}p} = -\int_0^\infty \frac{\partial}{\partial p}[u(t)\mathrm{e}^{-pt}]\,\mathrm{d}t = \int_0^\infty tu(t)\mathrm{e}^{-pt}\,\mathrm{d}t.$$

The differentiated integral converges uniformly in $\mathscr{R}p \geqslant A_1 > A$.
A detailed proof will be given in P3.1.1. This result may also be used to obtain E1.1.6.

**D1.1.3** The Heaviside *unit function is*

$$\mathrm{H}(t) = \begin{cases} 0 & (t<0), \\ 1 & (t>0). \end{cases}$$

With this notation the Laplace transform integral (1.1.1) may be written as

$$\bar{u}(p) = \int_{-\infty}^{\infty} u(t)\mathrm{H}(t)\mathrm{e}^{-pt}\,\mathrm{d}t, \tag{1.1.6}$$

and the factor $\mathrm{H}(t)$ represents a 'switching-on' of the function $u(t)$ at $t=0$. Equation (1.1.6) expresses $\bar{u}(p)$ as the *two-sided* Laplace transform of the function $u(t)\mathrm{H}(t)$.

**P1.1.10**

$$\mathscr{L}\{u(t-a)\mathrm{H}(t-a)\} = \mathrm{e}^{-ap}\bar{u}(p) \qquad \text{if } a \geqslant 0.$$

Since $\mathrm{H}(t-a)=0$ for $t<a$, the left side is

$$\int_a^\infty u(t-a)\mathrm{e}^{-pt}\,\mathrm{d}t = \int_0^\infty u(s)\mathrm{e}^{-p(s+a)}\,\mathrm{d}s = \mathrm{e}^{-ap}\bar{u}(p).$$

The condition $a \geqslant 0$ is necessary if we use the standard definition (1.1.1) of the Laplace transform. The factor $\mathrm{e}^{-ap}$ is the *delay operator* on the transform, since it corresponds to switching on the function $u(t)$ at $t=a$ instead of $t=0$, with the retarded argument $(t-a)$ in place of $t$.

**E1.1.10** Let $u(t)$ be the rectangular pulse function (shown in Figure 1.1) defined by

$$\begin{aligned} u(t) &= \begin{cases} 0 & (t<a), \\ k & (a<t<b), \\ 0 & (t>b), \end{cases} \\ &= k[\mathrm{H}(t-a)-\mathrm{H}(t-b)]. \end{aligned}$$

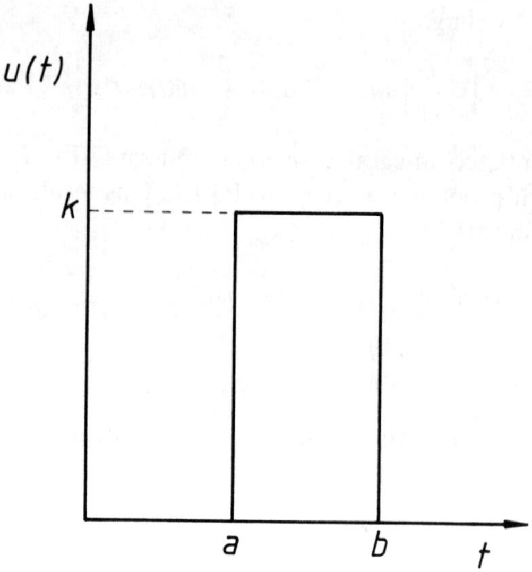


Figure 1.1

Then, from P1.1.10 or directly, we have

$$\bar{u}(p) = k(e^{-ap} - e^{-bp})p^{-1}.$$

Any step function is the sum of functions of this type (Figure 1.2).

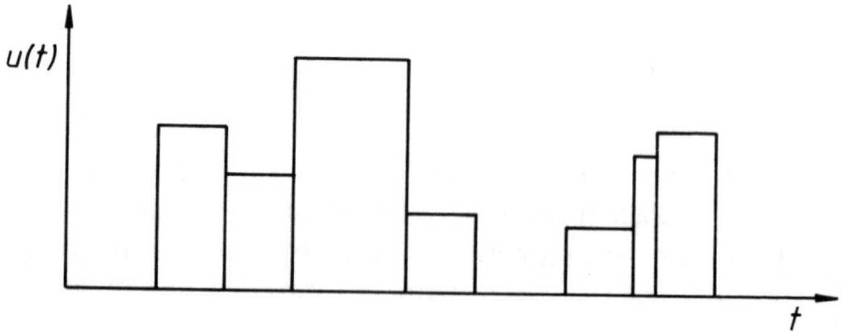


Figure 1.2

If we consider the limiting process in which $k \to \infty$ as $b \to a$ in such a way that $k(b-a) \to 1$, E1.1.10 gives a representation of the Dirac *delta function* (or impulse function) $\delta(t-a)$. This is not an ordinary function but a *generalized function.*

**D1.1.4**

$$\delta(t-a)=0 \qquad \text{if } t\neq a,$$

$$\int_{-\infty}^{\infty}\delta(t-a)f(t)\,\mathrm{d}t=f(a)$$

for any function $f(t)$ continuous at $t=a$.

If $u(t)$ in E1.1.10 represents a force, then $\delta(t-a)$ is an impulsive force; if $u(t)$ represents an electric current, then $\delta(t-a)$ is the instantaneous transport of a charge. Singularities of the type of a source can be represented by means of delta functions, and the solutions of partial differential equations, as given in Chapter 4, can often be interpreted as arising from source distributions.

The same limiting process applied to $\bar{u}(p)$ of E1.1.10 gives the transform of the delta function.

**E1.1.11**

$$\mathscr{L}\{\delta(t-a)\}=\lim_{b\to a}\frac{\mathrm{e}^{-ap}-\mathrm{e}^{-bp}}{(b-a)p}=\mathrm{e}^{-ap} \qquad (a\geqslant 0).$$

This also comes from D1.1.4 with $f(t)=\mathrm{e}^{-pt}$. In particular,

$$\mathscr{L}\{\delta(t)\}=1.$$

**P1.1.11** If $u(t)$ is a periodic function with period $T$ and

$$v(t)=u(t)\mathrm{H}(T-t)$$

then

$$\bar{u}(p)=(1-\mathrm{e}^{-pT})^{-1}\,\bar{v}(p).$$

Since $u(t)$ is periodic and $v(t)=0$ for $t>T$, the relation between $u(t)$ and $v(t)$ can be expressed as

$$u(t)=v(t)+u(t-T)\mathrm{H}(t-T),$$

as can be seen from Figure 1.3, so that from P1.1.10

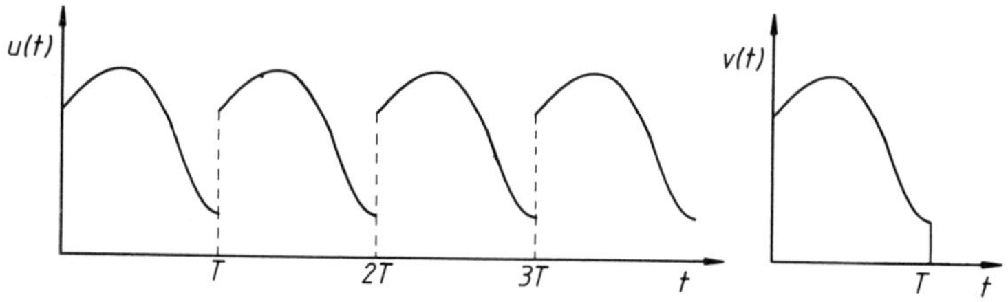


Figure 1.3

$$\bar{u}(p) = \bar{v}(p) + e^{-pT}\bar{u}(p).$$

Note that $\bar{v}(p)$ is defined for all $p$, and $\mathscr{L}\{u(t)\}$ converges for $\mathscr{R}p > 0$. The connexion between P1.1.11 and the Fourier series of $u(t)$ will be treated in Section 5.3.

**E1.1.12** The 'square-wave' function with period $T$ is defined by

$$u(t) = \begin{cases} 1 & (nT < t < (n+\tfrac{1}{2})T), \\ -1 & ((n+\tfrac{1}{2})T < t < (n+1)T), \end{cases}$$

for $n = 0, 1, 2 \ldots$ (see Figure 1.4). Here

$$v(t) = H(t) - 2H(t - \tfrac{1}{2}T) + H(t - T),$$

so that

$$\bar{v}(p) = (1 - 2e^{-pT\ 2} + e^{-pT})p^{-1}.$$

Hence we obtain

$$\bar{u}(p) = \frac{(1 - e^{-pT/2})^2}{(1 - e^{-pT})p} = \frac{1 - e^{-pT/2}}{1 + e^{-pT/2}}p^{-1}$$
$$= p^{-1} \tanh \tfrac{1}{4}pT.$$

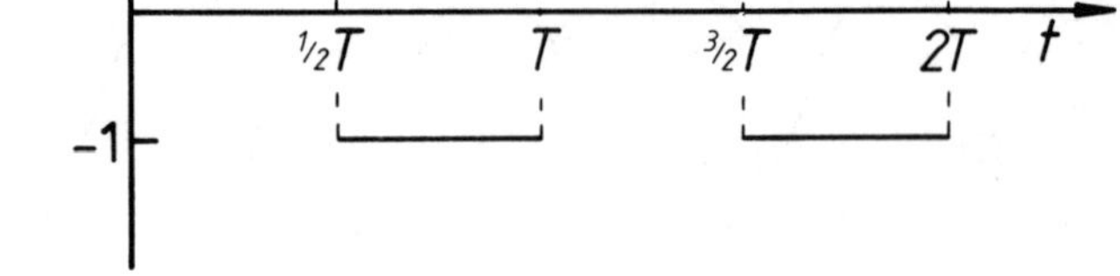

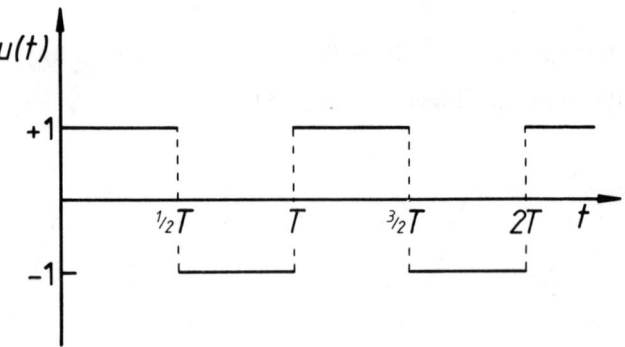

Figure 1.4

## 1.2. Further Properties

**D1.2.1** The *convolution* of two functions $u(t)$ and $v(t)$ defined for $t > 0$ is

$$u(t) * v(t) = \int_0^t u(s)v(t-s)\,\mathrm{d}s. \tag{1.2.1}$$

**P1.2.1**

(i) $u(t) * v(t) = v(t) * u(t)$,

(ii) $u(t) * [\lambda v(t) + \mu w(t)] = \lambda u(t) * v(t) + \mu u(t) * w(t)$,

(iii) $u(t) * [v(t) * w(t)] = [u(t) * v(t)] * w(t)$.

The substitution $s = t - r$ in (1.2.1) gives (i), and (ii) is obvious. For (iii) we have

$$\begin{aligned} u(t) * [v(t) * w(t)] &= \int_0^t u(s)\left( \int_0^{t-s} v(r)w(t-s-r)\,\mathrm{d}r \right)\mathrm{d}s \\ &= \int_0^t \left( \int_s^t u(s)v(x-s)w(t-x)\,\mathrm{d}x \right)\mathrm{d}s \\ &= \int_0^t \left( \int_0^x u(s)v(x-s)\,\mathrm{d}s \right) w(t-x)\,\mathrm{d}x \\ &= [u(t) * v(t)] * w(t), \end{aligned}$$

by putting $r = x - s$ and changing the order of integration.

P1.2.1 shows that convolution has properties analogous to multiplication.

**P1.2.2** If $w(t) = u(t) * v(t)$ then $\bar{w}(p) = \bar{u}(p)\bar{v}(p)$.

For

$$\begin{aligned} \bar{w}(p) &= \int_0^\infty \left( \int_0^t u(s)v(t-s)\,\mathrm{d}s \right) \mathrm{e}^{-pt}\,\mathrm{d}t \\ &= \int_0^\infty \left( \int_s^\infty u(s)v(t-s)\mathrm{e}^{-pt}\,\mathrm{d}t \right)\mathrm{d}s \\ &= \int_0^\infty u(s)\left( \int_0^\infty v(r)\mathrm{e}^{-p(r+s)}\,\mathrm{d}r \right)\mathrm{d}s = \bar{u}(p)\bar{v}(p), \end{aligned}$$

where the order of integration has been reversed and $t = r + s$.

**E1.2.1**

$$e^{\alpha t} * e^{\beta t} = \int_0^t e^{\alpha s} e^{\beta(t-s)}\, ds = e^{\beta t} \int_0^t e^{(\alpha-\beta)s}\, ds.$$

If $\alpha \neq \beta$, this is

$$e^{\beta t}\frac{e^{(\alpha-\beta)t}-1}{\alpha-\beta} = \frac{e^{\alpha t}-e^{\beta t}}{\alpha-\beta}.$$

Thus

$$\mathscr{L}\{e^{\alpha t} * e^{\beta t}\} = \frac{1}{\alpha-\beta}\left(\frac{1}{p-\alpha} - \frac{1}{p-\beta}\right) = \frac{1}{(p-\alpha)(p-\beta)}.$$

Also

$$e^{\alpha t} * e^{\alpha t} = \int_0^t e^{\alpha s} e^{\alpha(t-s)}\, ds = \int_0^t e^{\alpha t}\, ds = te^{\alpha t},$$

so that

$$\mathscr{L}\{e^{\alpha t} * e^{\alpha t}\} = (p-\alpha)^{-2}.$$

Further examples of convolutions are given as E1.4.2–E1.4.4 to illustrate the use of P1.2.2 in working out inverse transforms.

In the analogy between convolution and multiplication, the unit for convolution is the delta function, since

$$u(t) * \delta(t) = \int_0^t u(s)\delta(t-s)\, ds = u(t),$$

and P1.2.2 corresponds to $\mathscr{L}\{\delta(t)\} = 1$. Note also that

$$u(t) * 1 = \int_0^t u(s)\, ds,$$

and in this case P1.2.2 reduces to P1.1.6.

If the functions $u(t)$ and $v(t)$ are defined for all real $t$, the corresponding definition of convolution is

$$u(t) * v(t) = \int_{-\infty}^{\infty} u(s)v(t-s)\, ds,$$

which reduces to equation (1.2.1) when the functions $u(t)$ and $v(t)$ are multiplied by the unit function $H(t)$.

The arguments given above are purely formal; to justify them it will be shown that if $u(t)$ and $v(t)$ satisfy Assumptions 1 and 2, then so does $u(t) * v(t)$. We need a lemma.

**P1.2.3** If $f(t)$ is absolutely integrable in $A \leqslant t \leqslant B$ and $A < a < b < B$, then

$$\int_a^b |f(t+\eta) - f(t)| \, dt \to 0 \qquad \text{as } \eta \to 0.$$

Given any $\varepsilon > 0$, there is a step function

$$\phi(t) = \sum_{n=1}^{N} k_n [H(t - t_{n-1}) - H(t - t_n)],$$

where $A = t_0 < t_1 < \ldots < t_N = B$, such that

$$\int_A^B |f(t) - \phi(t)| \, dt < \varepsilon.$$

Then

$$\int_a^b |f(t+\eta) - f(t)| \, dt \leqslant \int_a^b |f(t+\eta) - \phi(t+\eta)|' dt$$

$$+ \int_a^b |f(t) - \phi(t)| \, dt + \int_a^b |\phi(t+\eta) - \phi(t)|' dt$$

$$< 2\varepsilon + \sum_{n=1}^{N} |k_n - k_{n-1}| |\eta|,$$

provided that $a + \eta > A$, $b + \eta < B$ and $|\eta| < \min(t_n - t_{n-1})$. Hence

$$\int_a^b |f(t+\eta) - f(t)| \, dt < 3\varepsilon$$

if also

$$|\eta| < \varepsilon \left( \sum_{n=1}^{N} |k_n - k_{n-1}| \right)^{-1}$$

**P.1.2.4** Suppose that $u(t)$ and $v(t)$ are integrable in $0 \leqslant t \leqslant T$ with the exception of the neighbourhoods of $t = a_1, a_2, \ldots, a_m$ for $u(t)$ and $t = b_1, b_2, \ldots, b_n$ for $v(t)$, and that

$$\int_0^T |u(t)| \, dt \qquad \text{and} \qquad \int_0^T |v(t)| \, dt$$

both exist. Then

$$w(t) = \int_0^t u(s) v(t-s) \, ds$$

is continuous at $t = \tau$ provided that $0 < \tau < T$ and $\tau \neq a_i + b_j$ for any $i, j$.

$u(s)$ and $v(\tau - s)$ are bounded except in the neighbourhoods of the distinct points $s = a_1, a_2, \ldots, a_m, \tau - b_1, \tau - b_2, \ldots, \tau - b_n$. Define $v(t) = 0$ for $t < 0$. If $0 < h < \min(|\tau - a_i - b_j|, T - \tau)$, we can enclose $a_1, \ldots, a_m$ in a set of intervals $I$ such that $v(\tau + \eta - s)$ is bounded, say $|v(\tau + \eta - s)| < M$, whenever $|\eta| \leqslant h$ and $s$ belongs to $I$. Also, we can find a set of intervals $J$, contained in $I$ but still containing $a_1, \ldots, a_m$, such that

$$\int_J |u(s)| \, ds < \varepsilon / M.$$

Now

$$w(\tau + \eta) - w(\tau) = \int_0^{\tau + h} u(s)[v(\tau + \eta - s) - v(\tau - s)] \, ds,$$

and $u(s)$ is bounded in $K$, the remainder of $0 \leqslant s \leqslant \tau + h$ after $J$ has been removed, say $|u(s)| < N$ in $K$. Thus

$$\begin{aligned} |w(\tau + \eta) - w(\tau)| &\leqslant \int_0^{\tau + h} |u(s)| \, |v(\tau + \eta - s) - v(\tau - s)| \, ds \\ &< \int_J |u(s)| . 2M \, ds \\ &\quad + \int_K N |v(\tau + \eta - s) - v(\tau - s)| \, ds \\ &< 2\varepsilon + N \int_0^{\tau + h} |v(\tau + \eta - s) - v(\tau - s)| \, ds < 3\varepsilon \end{aligned}$$

if $|\eta|$ is sufficiently small, from P1.2.3.

**P1.2.5** With the hypotheses of P1.2.4, $w(t) \to 0$ as $t \to 0+$ unless $a_1 = b_1 = 0$.

If $a_1 > 0$ and $b_1 > 0$, both $u(t)$ and $v(t)$ are bounded near $t = 0$. If $a_1 = 0$ and $b_1 > 0$, then for $0 < t < T_1 < \min(a_2, b_1)$

$$w(t) = \lim_{\delta \to 0} \int_\delta^t u(s) v(t - s) \, ds,$$

and $|v(t - s)| < M$ (say), so that

$$|w(t)| \leqslant M \int_0^t |u(s)| \, ds \to 0 \qquad \text{as } t \to 0+.$$

**P1.2.6** With the hypotheses of P1.2.4,

$$\int_0^T |w(t)|\,dt \leqslant \int_0^T \left( \int_0^t |u(s)|\,|v(t-s)|\,ds \right) dt$$
$$\leqslant \int_0^T |u(s)|\,ds \int_0^T |v(r)|\,dr,$$

on putting $t = r + s$.

**P1.2.7** If

$$\int_0^\infty |u(t)|\,e^{-xt}\,dt \qquad \text{and} \qquad \int_0^\infty |v(t)|\,e^{-xt}\,dt$$

both converge for some real $x$, then so does

$$\int_0^\infty |w(t)|\,e^{-xt}\,dt.$$

As in P1.2.6, for any $T > 0$

$$\int_0^T |w(t)|\,e^{-xt}\,dt \leqslant \int_0^T |u(s)|\,e^{-xs}\,ds \int_0^T |v(r)|\,e^{-xr}\,dr.$$

Thus $w(t)$ satisfies Assumptions 1 and 2, and P1.2.1 and P1.2.2 are justified by absolute convergence.

**D1.2.2** The *gamma function* is defined by

$$\Gamma(v) = \int_0^\infty x^{v-1} e^{-x}\,dx \qquad \text{for } \mathscr{R}v > 0.$$

**E1.2.2**

$$\mathscr{L}\{t^{v-1}\} = \Gamma(v)p^{-v} \qquad \text{for } \mathscr{R}v > 0, \mathscr{R}p > 0.$$

If $p$ is real, the substitution $pt = x$ gives the result. If $p$ is complex, it is necessary to justify rotating the path of integration in a complex $x$ plane. Comparing E1.2.2 with E1.1.6, we see that $\Gamma(n) = (n-1)!$ for $n = 1, 2, 3, \ldots$ From either P1.1.6 or P1.1.9, we find that

$$\Gamma(v+1) = v\Gamma(v) \qquad \text{for } \mathscr{R}v > 0,$$

and this functional equation may be used to define $\Gamma(v)$ for all complex $v$ except $0, -1, -2, \ldots$ in the form

$$\Gamma(v) = \frac{\Gamma(v+N)}{v(v+1)\ldots(v+N-1)} \qquad \text{for } \mathscr{R}v > -N,$$

where $N$ is any positive integer. The notation $\nu! = \Gamma(\nu+1)$ is also used for general values of $\nu$.

**E1.2.3**

$$t^{\mu-1} * t^{\nu-1} = \int_0^t s^{\mu-1}(t-s)^{\nu-1}\,\mathrm{d}s = \mathrm{B}(\mu,\nu)t^{\mu+\nu-1}$$

on putting $s = tx$, where the *beta function* is defined as follows.

**D1.2.3**

$$\mathrm{B}(\mu,\nu) = \int_0^1 x^{\mu-1}(1-x)^{\nu-1}\,\mathrm{d}x \qquad \text{for } \mathscr{R}\mu > 0,\ \mathscr{R}\nu > 0.$$

From P1.2.2 and E1.2.2, we have

$$\Gamma(\mu)p^{-\mu}\cdot\Gamma(\nu)p^{-\nu} = \mathrm{B}(\mu,\nu)\Gamma(\mu+\nu)p^{-\mu-\nu},$$

so that

$$\mathrm{B}(\mu,\nu) = \frac{\Gamma(\mu)\Gamma(\nu)}{\Gamma(\mu+\nu)}.$$

In particular, the substitution $x = \sin^2\theta$ in D1.2.3 gives $\mathrm{B}(\frac{1}{2},\frac{1}{2}) = \pi$, and since $\Gamma(1) = 1$ and $\Gamma(\frac{1}{2}) > 0$ we obtain $\Gamma(\frac{1}{2}) = \sqrt{\pi}$.

The remaining properties to be treated in this section relate to the behaviour of the transform $\bar{u}(p)$ as $p \to \infty$ or $p \to 0$,

**P1.2.8** $\bar{u}(p) \to 0$ uniformly as $\mathscr{R}p \to +\infty$.

Let $p = x + \mathrm{i}y$ where $x \geqslant A$, and let

$$\int_0^\infty |u(t)|\mathrm{e}^{-At}\,\mathrm{d}t = M.$$

Then

$$\left|\int_0^\infty u(t)\mathrm{e}^{-pt}\,\mathrm{d}t\right| \leqslant \int_0^\infty |u(t)|\mathrm{e}^{-xt}\,\mathrm{d}t = \left(\int_0^\tau + \int_\tau^\infty\right)|u(t)|\mathrm{e}^{-xt}\,\mathrm{d}t.$$

Now

$$\int_0^\tau |u(t)|\mathrm{e}^{-xt}\,\mathrm{d}t \leqslant \int_0^\tau |u(t)|\mathrm{e}^{-At}\,\mathrm{d}t < \varepsilon$$

if we choose $\tau$ sufficiently small, and

$$\int_\tau^\infty |u(t)|\mathrm{e}^{-xt}\,\mathrm{d}t = \int_\tau^\infty |u(t)|\mathrm{e}^{-At}\cdot\mathrm{e}^{-(x-A)t}\,\mathrm{d}t$$
$$\leqslant \mathrm{e}^{-(x-A)\tau}\int_\tau^\infty |u(t)|\mathrm{e}^{-At}\,\mathrm{d}t$$
$$\leqslant M\mathrm{e}^{-(x-A)\tau}.$$

Thus $|\bar{u}(p)| < 2\varepsilon$ if $x > A + \tau^{-1}\log(M/\varepsilon)$.

Note that P1.2.8 does not apply to $u(t) = \delta(t)$ (E1.1.11), which is not an ordinary function and does not satisfy the standard assumptions.

**P1.2.9** If $u(t) \to l$ as $t \to 0+$, then $p\bar{u}(p) \to l$ as $|p| \to \infty$ with $|\arg p| \leqslant \frac{1}{2}\pi - \delta < \frac{1}{2}\pi$.

**P1.2.10** If $u(t) \to l$ as $t \to \infty$, then $p\bar{u}(p) \to l$ as $|p| \to 0$ with $|\arg p| \leqslant \frac{1}{2}\pi - \delta < \frac{1}{2}\pi$.

For the case of an e.t.f. $u(t)$ P1.2.9 was proved in E1.1.7, and P1.2.10 can be obtained similarly, without any restriction on $\arg p$. In general, the implication is only as stated in these properties. Since $\mathscr{L}\{1\} = p^{-1}$, they follow from the corresponding propositions with $l = 0$, which will now be proved.

**P1.2.11** If $u(t) \to 0$ as $t \to 0+$, then $p\bar{u}(p) \to 0$ as $|p| \to \infty$ with $|\arg p| \leqslant \frac{1}{2}\pi - \delta$.

$u(t)$ is bounded in some interval, say $|u(t)| < K$ for $0 < t \leqslant T$. We can suppose that in Assumption 2 $A > 0$. Then

$$\int_T^\infty u(t)\mathrm{e}^{-pt}\mathrm{d}t = [v(t)\mathrm{e}^{-(p-A)t}]_T^\infty + (p-A)\int_T^\infty v(t)\mathrm{e}^{-(p-A)t}\,\mathrm{d}t,$$

where

$$v(t) = \int_0^t u(s)\mathrm{e}^{-As}\mathrm{d}s.$$

Since

$$|v(t)| \leqslant M = \int_0^\infty |u(s)|\mathrm{e}^{-As}\mathrm{d}s,$$

$$\left|p\int_T^\infty u(t)\mathrm{e}^{-pt}\mathrm{d}t\right| \leqslant M|p|\mathrm{e}^{-(x-A)T} + \frac{|p|(|p|+A)}{x-A}M\mathrm{e}^{-(x-A)T}$$

for $\mathscr{R}p = x > A$. Since $|p| \leqslant x \operatorname{cosec} \delta$, we shall have

$$\left| p \int_T^\infty u(t) \mathrm{e}^{-pt} \mathrm{d}t \right| < \varepsilon \qquad \text{when } x > X_0(\varepsilon).$$

Let $m(\tau)$ be the least upper bound of $|u(t)|$ in $0 < t \leqslant \tau$, so that $m(\tau)$ decreases with $\tau$ and $m(\tau) \to 0$ as $\tau \to 0+$. Then

$$\left| p \int_0^\tau u(t) \mathrm{e}^{-pt} \mathrm{d}t \right| \leqslant |p| \tau m(\tau) \leqslant \varepsilon$$

if $x\tau m(\tau) \leqslant \varepsilon \sin \delta$. Also

$$\left| p \int_\tau^T u(t) \mathrm{e}^{-pt} \mathrm{d}t \right| \leqslant K |p| \int_\tau^T e^{-xt} \mathrm{d}t \leqslant K(\operatorname{cosec} \delta) \mathrm{e}^{-x\tau} \leqslant \varepsilon$$

if $x\tau \geqslant \log(\varepsilon^{-1} K \operatorname{cosec} \delta)$.

Choose $\tau_1$ such that

$$m(\tau_1) \leqslant \frac{\varepsilon \sin \delta}{\log(\varepsilon^{-1} K \operatorname{cosec} \delta)}$$

and put

$$X_1(\varepsilon) = \max(X_0(\varepsilon),\ \tau_1^{-1} \log(\varepsilon^{-1} K \operatorname{cosec} \delta)).$$

For any $x > X_1(\varepsilon)$, take $\tau = x^{-1} \log(\varepsilon^{-1} K \operatorname{cosec} \delta)$. Then $\tau \leqslant \tau_1$ so that $m(\tau) \leqslant m(\tau_1)$. Thus

$$|p\bar{u}(p)| < 3\varepsilon \qquad \text{for } |p| > X_1(\varepsilon) \operatorname{cosec} \delta,\ |\arg p| \leqslant \tfrac{1}{2}\pi - \delta.$$

**P1.2.12** If $u(t) \to 0$ as $t \to \infty$, then $p\bar{u}(p) \to 0$ as $|p| \to 0$ with $|\arg p| \leqslant \frac{1}{2}\pi - \delta$.

Since $u(t)$ is bounded for large $t$, $\bar{u}(p)$ converges absolutely if $\mathscr{R}p > 0$. Choose $T$ such that $|u(t)| < \varepsilon \sin \delta$ for $t \geqslant T$. Then if $\mathscr{R}p = x > 0$

$$\left| p \int_T^\infty u(t) \mathrm{e}^{-pt} \mathrm{d}t \right| \leqslant |p| \varepsilon \sin \delta \int_T^\infty \mathrm{e}^{-xt} \mathrm{d}t \leqslant \varepsilon.$$

Also

$$\left| p \int_0^T u(t) \mathrm{e}^{-pt} \mathrm{d}t \right| \leqslant |p| \int_0^T |u(t)| \mathrm{d}t,$$

so that $|p\bar{u}(p)| < 2\varepsilon$ if

$$|p| < \varepsilon\left(\int_0^T |u(t)|\,\mathrm{d}t\right)^{-1} \qquad \text{and} \qquad |\arg p| \leqslant \tfrac{1}{2}\pi - \delta.$$

The limit formula P1.2.9 can be generalized if we have a series expansion for $u(t)$ about $t = 0$. If

$$u(t) = \sum_{n=0}^{\infty} a_n t^{n\mu+\nu-1}$$

and we transform the series term by term using E1.2.2, we get

$$\bar{u}(p) \sim \sum_{n=0}^{\infty} a_n \frac{\Gamma(n\mu+\nu)}{p^{n\mu+\nu}}.$$

In general, this series is not convergent but *asymptotic*, as indicated by the symbol $\sim$. This means that for each $N = 0, 1, 2, \ldots$

$$p^{N\mu+\nu}\left(\bar{u}(p) - \sum_{n=0}^{N} a_n \frac{\Gamma(n\mu+\nu)}{p^{n\mu+\nu}}\right) \to 0 \qquad \text{as } |p| \to \infty.$$

The result is known as *Watson's lemma* (Watson, 1944, Section 8.3). We shall prove it in the following form.

**P1.2.13** If

$$u(t) \sim \sum_{n=0}^{\infty} a_n t^{n\mu+\nu-1} \qquad \text{as } t \to 0+,$$

where $\mathscr{R}\mu > 0$, $\mathscr{R}\nu > 0$, then

$$\bar{u}(p) \sim \sum_{n=0}^{\infty} a_n \Gamma(n\mu+\nu) p^{-n\mu-\nu}$$

as $|p| \to \infty$ with $|\arg p| \leqslant \frac{1}{2}\pi - \delta < \frac{1}{2}\pi$.

Put $\mu = \mu_1 + \mathrm{i}\mu_2$, $\nu = \nu_1 + \mathrm{i}\nu_2$ and let

$$u_N(t) = u(t) - \sum_{n=0}^{N} a_n t^{n\mu+\nu-1}.$$

Then $|u_N(t)| < Kt^{\lambda}$ for $0 < t \leqslant T$ (say), where $\lambda = (N+1)\mu_1 + \nu_1 - 1$. Also, if

$$\int_0^{\infty} |u(t)| \mathrm{e}^{-At}\,\mathrm{d}t$$

converges, where $A > 0$, then

$$\int_0^\infty |u_N(t)| e^{-At} dt = M$$

is finite. Let $\mathscr{R} p = x > A$. Then

$$|\bar{u}_N(p)| \leqslant \int_0^T |u_N(t)| e^{-xt} dt + \int_T^\infty |u_N(t)| e^{-xt} dt$$

$$\leqslant \int_0^T K t^\lambda e^{-xt} dt + \int_T^\infty |u_N(t)| e^{-At} . e^{-(x-A)t} dt$$

$$\leqslant K \int_0^\infty t^\lambda e^{-xt} dt + e^{-(x-A)T} \int_T^\infty |u_N(t)| e^{-At} dt$$

$$\leqslant K\Gamma(\lambda + 1) x^{-(\lambda+1)} + M e^{-(x-A)T} \leqslant L x^{-(\lambda+1)},$$

where $L$ is independent of $x$. Since $|\arg p| \leqslant \frac{1}{2}\pi - \delta$

$$|p^{N\mu+\nu} \bar{u}_N(p)| \leqslant \frac{L \exp(\frac{1}{2}\pi |N\mu_2 + \nu_2|)}{|p|^{\mu_1} \sin^{\lambda+1}\delta} \to 0 \qquad \text{as } |p| \to \infty .$$

If $\nu = 1$, the case $N = 0$ gives P1.2.9, but in the proof of P1.2.11 nothing was assumed about the way the limit was approached as $t \to 0+$.

**E1.2.4**

$$u(t) = e^{\alpha t} = \sum_{n=0}^\infty (\alpha t)^n / n!$$

P1.2.13 gives

$$\bar{u}(p) \sim \sum_{n=0}^\infty \frac{\alpha^n}{n!} \cdot \frac{n!}{p^{n+1}} = p^{-1} \sum_{n=0}^\infty \left(\frac{\alpha}{p}\right)^n .$$

This series converges to the exact result $(p - \alpha)^{-1}$ if $|p| > |\alpha|$.

**E1.2.5**

$$u(t) = \exp(-t^2) = \sum_{n=0}^\infty \frac{(-1)^n}{n!} t^{2n} .$$

Hence

$$\bar{u}(p) \sim \sum_{n=0}^{\infty} \frac{(-1)^n}{n!} \frac{(2n)!}{p^{2n+1}} = p^{-1} \sum_{n=0}^{\infty} (n-\tfrac{1}{2})(n-\tfrac{3}{2})\ldots\tfrac{1}{2}\left(-\frac{4}{p^2}\right)^n.$$

This series does not converge for any finite $p$, so that the result is only asymptotic.

### 1.3 Methods of Evaluation

We have seen that the substitution of a power series for $u(t)$ in the transform integral (1.1.1) does not always give a convergent series for $\bar{u}(p)$. A simple condition for the transformed series to converge to $\bar{u}(p)$ is the following.

**P1.3.1** If

$$u(t) = \sum_{n=0}^{\infty} a_n t^{n+\nu-1},$$

where $\mathscr{R}\nu > 0$ and $|a_n| \leqslant kA^n/n!$, then

$$\bar{u}(p) = \sum_{n=0}^{\infty} a_n \Gamma(n+\nu) p^{-n-\nu} \qquad \text{for } \mathscr{R}p > A.$$

The power series converges for all $t$, so that $u(t)$ is continuous for $t > 0$. Let

$$u_N(t) = u(t) - \sum_{n=0}^{N} a_n t^{n+\nu-1}.$$

Then if $\nu = \nu_1 + i\nu_2$

$$|u_N(t)| \leqslant kt^{\nu_1 - 1} \sum_{n=N+1}^{\infty} \frac{(At)^n}{n} = kt^{\nu_1 - 1}\left(e^{At} - \sum_{n=0}^{N} \frac{(At)^n}{n!}\right)$$

for all $t > 0$. Hence if $\mathscr{R}p = x > A$

$$|\bar{u}_N(p)| \leqslant k \frac{\Gamma(\nu_1)}{(x-A)^{\nu_1}} - k \sum_{n=0}^{N} \frac{\Gamma(n+\nu_1)A^n}{n!\,x^{n+\nu_1}}$$

so that $|\bar{u}_N(p)| \to 0$ as $N \to \infty$.

**D1.3.1** *Bessel functions.* These are defined by

$$\mathrm{J}_\nu(t) = \sum_{r=0}^{\infty} \frac{(-1)^r(\frac{1}{2}t)^{\nu+2r}}{r!\Gamma(\nu+r+1)},$$

$$\mathrm{Y}_\nu(t) = \mathrm{J}_\nu(t)\cot \nu\pi - \mathrm{J}_{-\nu}(t)\,\mathrm{cosec}\,\nu\pi,$$

$$\mathrm{I}_\nu(t) = \mathrm{i}^{-\nu}\mathrm{J}_\nu(\mathrm{i}t) = \sum_{r=0}^{\infty} \frac{(\frac{1}{2}t)^{\nu+2r}}{r!\Gamma(\nu+r+1)},$$

$$\mathrm{K}_\nu(t) = \tfrac{1}{2}\pi\,\mathrm{cosec}\,\nu\pi\,[\mathrm{I}_{-\nu}(t) - \mathrm{I}_\nu(t)].$$

$\mathrm{J}_\nu(t)$ and $\mathrm{Y}_\nu(t)$ are independent solutions of *Bessel's equation*

$$t^2\frac{\mathrm{d}^2u}{\mathrm{d}t^2} + t\frac{\mathrm{d}u}{\mathrm{d}t} + (t^2 - \nu^2)u = 0;$$

$\mathrm{I}_\nu(t)$ and $\mathrm{K}_\nu(t)$ are independent solutions of the modified equation

$$t^2\frac{\mathrm{d}^2u}{\mathrm{d}t^2} + t\frac{\mathrm{d}u}{\mathrm{d}t} - (t^2 + \nu^2)u = 0.$$

If $n$ is an integer, the definitions of $\mathrm{Y}_n(t)$ and $\mathrm{K}_n(t)$ fail, so they are defined as the limits of $\mathrm{Y}_\nu(t)$ and $\mathrm{K}_\nu(t)$ as $\nu \to n$. Details are given by Watson (1944, Chapter 3).

**E1.3.1** If $\mathscr{R}\nu > -1$ and $\mathscr{R}p > 0$ then

$$\mathscr{L}\{t^{\nu/2}\mathrm{J}_\nu(2\sqrt{(\alpha t)})\} = \mathscr{L}\left\{\sum_{r=0}^{\infty} \frac{(-1)^r\alpha^{(\nu/2)+r}t^{\nu+r}}{r!\Gamma(\nu+r+1)}\right\}$$

$$= \sum_{r=0}^{\infty} \frac{(-1)^r\alpha^{(\nu/2)+r}}{p^{\nu+r+1}} = \alpha^{\nu/2}p^{-\nu-1}\exp\left(-\frac{\alpha}{p}\right).$$

**E1.3.2** Similarly, if $\mathscr{R}\nu > -1$ and $\mathscr{R}p > |\alpha|$

$$\mathscr{L}\{\mathrm{J}_\nu(\alpha t)\} = (\tfrac{1}{2}\alpha)^\nu p^{-\nu-1}\sum_{r=0}^{\infty} \frac{\Gamma(\nu+2r+1)}{r!\Gamma(\nu+r+1)}\left(-\frac{\alpha^2}{4p^2}\right)^r.$$

When $\nu = 0$, this is

$$\mathscr{L}\{\mathrm{J}_\nu(\alpha t)\} = p^{-1}\sum_{r=0}^{\infty} \frac{(2r)!}{(r!)^2}\left(-\frac{\alpha^2}{4p^2}\right)^r$$

$$= p^{-1}\sum_{r=0}^{\infty} \frac{(r-\frac{1}{2})(r-\frac{3}{2})\ldots\frac{1}{2}}{r!}\left(-\frac{\alpha^2}{p^2}\right)^r$$

$$= p^{-1}\left(1 + \frac{\alpha^2}{p^2}\right)^{-1/2} = (p^2 + \alpha^2)^{-1/2}.$$

In the general case, the series can be summed to give

$$\mathscr{L}\{J_\nu(\alpha t)\} = (p^2 + \alpha^2)^{-1/2}\left(\frac{\alpha}{(p^2+\alpha^2)^{1/2}+p}\right)^\nu,$$

but this result is more easily derived from Bessel's differential equation, as in E2.4.3.

Another useful method for obtaining Laplace transforms is differentiation or integration with respect to a parameter.

**P1.3.2** If $v(t,\lambda) = \partial u(t,\lambda)/\partial\lambda$ is absolutely integrable, as a function of both variables, and its Laplace transform $\bar{v}(p,\lambda)$ converges uniformly with respect to $\lambda$ in $\lambda_1 \leqslant \lambda \leqslant \lambda_2$, then

(i) $\bar{u}(p,\lambda_2) - \bar{u}(p,\lambda_1) = \displaystyle\int_{\lambda_1}^{\lambda_2} \bar{v}(p,\lambda)\,\mathrm{d}\lambda,$

(ii) $\bar{v}(p,\lambda) = \partial\bar{u}(p,\lambda)/\partial\lambda \qquad$ for $\lambda_1 < \lambda < \lambda_2$.

By hypothesis, we can choose $T_0$ such that

$$\left|\int_T^\infty v(t,\lambda)\mathrm{e}^{-pt}\,\mathrm{d}t\right| < \varepsilon$$

for all $T \geqslant T_0$ and $\lambda_1 \leqslant \lambda \leqslant \lambda_2$. Since $v(t,\lambda)$ is absolutely integrable,

$$\left|\int_0^T\left(\int_{\lambda_1}^{\lambda_2} v(t,\lambda)\mathrm{e}^{-pt}\,\mathrm{d}\lambda\right)\mathrm{d}t - \int_{\lambda_1}^{\lambda_2}\left(\int_0^\infty v(t,\lambda)\mathrm{e}^{-pt}\,\mathrm{d}t\right)\mathrm{d}\lambda\right|$$

$$= \left|\int_{\lambda_1}^{\lambda_2}\left(\int_T^\infty v(t,\lambda)\mathrm{e}^{-pt}\,\mathrm{d}t\right)\mathrm{d}\lambda\right| < (\lambda_2 - \lambda_1)\varepsilon$$

for all $T \geqslant T_0$. When $T \to \infty$ we get (i). In (i), we can replace $\lambda_1, \lambda_2$ by any $\lambda_1', \lambda_2'$ such that $\lambda_1 < \lambda_1' < \lambda_2' < \lambda_2$, so that (ii) follows.

**E1.3.3** Since $\partial(t^{\nu-1})/\partial\nu = t^{\nu-1}\log t$, E1.2.2 gives

$$\mathscr{L}\{t^{\nu-1}\log t\} = \Gamma'(\nu)p^{-\nu} - \Gamma(\nu)p^{-\nu}\log p$$

for $\mathscr{R}p > 0$, $\mathscr{R}\nu > 0$. This result can also be expressed in terms of the *psi function*.

**D1.3.2**

$$\psi(\nu) = \frac{\Gamma'(\nu)}{\Gamma(\nu)} = \frac{\mathrm{d}}{\mathrm{d}\nu}[\log\Gamma(\nu)].$$

Thus

$$\mathscr{L}\{t^{\nu-1}\log t\} = [\psi(\nu) - \log p]\Gamma(\nu)p^{-\nu}.$$

In particular

$$\mathscr{L}\{\log t\} = -(\log p + \gamma)p^{-1},$$

where

$$\gamma = -\psi(1) = -\Gamma'(1) = -\int_0^\infty \log t\,\mathrm{e}^{-t}\mathrm{d}t$$

is *Euler's constant.*

**E1.3.4** $u(t, \lambda) = t^{-1}\sin\lambda t$ gives $v(t, \lambda) = \cos\lambda t$. Since $\bar{v}(p, \lambda) = p/(p^2 + \lambda^2)$ and $u(t, 0) = 0$, we find

$$\mathscr{L}\{t^{-1}\sin\lambda t\} = \int_0^\lambda \frac{p}{p^2+\mu^2}\,\mathrm{d}\mu = \tan^{-1}\left(\frac{\lambda}{p}\right).$$

**E1.3.5** Similarly,

$$\mathscr{L}\{t^{-1}(\cos\lambda t - \cos\mu t)\} = \tfrac{1}{2}\log\left(\frac{p^2+\mu^2}{p^2+\lambda^2}\right).$$

Note that if given a function such as $t^{-1}\sin t$ it is convenient to introduce a parameter $\lambda$ in order to differentiate with respect to $\lambda$. The last two examples can also be treated by the converse of P1.1.9.

**P1.3.3** If $v(t) = t^{-1}u(t)$ has a Laplace transform, it is given by

$$\bar{v}(p) = \int_p^\infty \bar{u}(z)\,\mathrm{d}z.$$

For $\mathrm{d}\bar{v}/\mathrm{d}p = -\bar{u}(p)$ and from P1.2.8 $\bar{v}(p) \to 0$ as $\mathscr{R}p \to \infty$.

**E1.3.4** Since $\mathscr{L}\{\sin\lambda t\} = \lambda/(p^2 + \lambda^2)$,

$$\mathscr{L}\{t^{-1}\sin\lambda t\} = \int_p^\infty \frac{\lambda}{z^2+\lambda^2}\,\mathrm{d}z = \left[\tan^{-1}\frac{z}{\lambda}\right]_p^\infty = \tfrac{1}{2}\pi - \tan^{-1}\left(\frac{p}{\lambda}\right).$$

The result of E1.3.5 can be obtained similarly.

In many cases the transform integral (1.1.1) cannot be expressed in terms of elementary functions, and the best that can be done is to reduce the integral to a standard form. Note that this may involve rotating the path of integration in a complex plane (see E1.2.2).

**E1.3.6** $u(t) = \exp(-t^2)$ has a Laplace transform for all $p$, and

$$\begin{aligned}\bar{u}(p) &= \int_0^\infty \exp[-(t+\tfrac{1}{2}p)^2]\exp(\tfrac{1}{4}p^2)\,\mathrm{d}t \\ &= \exp(\tfrac{1}{4}p^2)\int_{p/2}^\infty \exp(-s^2)\,\mathrm{d}s \\ &= \tfrac{1}{2}\sqrt{(\pi)}\exp(\tfrac{1}{4}p^2)\operatorname{erfc}(\tfrac{1}{2}p).\end{aligned}$$

Here erfc denotes the complementary *error function*.

**D1.3.3**

$$\operatorname{erf}(x) = \frac{2}{\sqrt{\pi}}\int_0^x \exp(-y^2)\,\mathrm{d}y,$$

$$\operatorname{erfc}(x) = \frac{2}{\sqrt{\pi}}\int_x^\infty \exp(-y^2)\,\mathrm{d}y = 1 - \operatorname{erf}(x).$$

**E1.3.7** $u(t) = (t+a)^{-1}$ has a transform for $\mathscr{R}p > 0$, if $a > 0$ or $a$ is complex. With $s = p(t+a)$

$$\bar{u}(p) = \int_{ap}^\infty \mathrm{e}^{ap-s}\frac{\mathrm{d}s}{s} = -\mathrm{e}^{ap}\operatorname{Ei}(-ap) = \mathrm{e}^{ap}E_1(ap).$$

Here $\operatorname{Ei}(x)$ and $E_1(x)$ are *exponential integrals.*

**D1.3.4**

$$\operatorname{Ei}(x) = \int_{-\infty}^x y^{-1}\mathrm{e}^y\,\mathrm{d}y,$$

$$E_n(x) = \int_1^\infty y^{-n}\mathrm{e}^{-xy}\,\mathrm{d}y$$

The result can also be obtained by taking the transform of the equation $(t+a)u(t) = 1$, namely

$$-\frac{\mathrm{d}\bar{u}}{\mathrm{d}p} + a\bar{u} = p^{-1}.$$

This differential equation for $\bar{u}(p)$ must be solved subject to the condition (P1.2.8) that $\bar{u} \to 0$ as $\mathscr{R}p \to \infty$.

**E1.3.8** This method also works for the function $u(t) = (t^2 + a^2)^{-1}$. The corresponding differential equation is

$$\frac{\mathrm{d}^2\bar{u}}{\mathrm{d}p^2} + a^2\bar{u} = p^{-1}.$$

Its general solution, obtained by variation of parameters, is

$$\bar{u}(p) = a^{-1}[A - \mathrm{Si}(ap)]\cos ap + a^{-1}[B + \mathrm{Ci}(ap)]\sin ap,$$

where $A$ and $B$ are constants and the *sine integral* Si and *cosine integral* Ci are defined as follows.

**D1.3.5**

$$\mathrm{Si}(x) = \int_0^x y^{-1}\sin y\,\mathrm{d}y,$$

$$\mathrm{Ci}(x) = -\int_x^\infty y^{-1}\cos y\,\mathrm{d}y.$$

Since $\mathrm{Si}(\infty) = \frac{1}{2}\pi$ (see E1.5.31), the required transform is

$$\mathscr{L}\{(t^2 + a^2)^{-1}\} = a^{-1}[\tfrac{1}{2}\pi - \mathrm{Si}(ap)]\cos ap + a^{-1}\mathrm{Ci}(ap)\sin ap.$$

**E1.3.9** $u(t) = t^\nu \exp(-\alpha/t)$ has a Laplace transform for all $\nu$ when $\mathscr{R}\alpha > 0$, $\mathscr{R}p > 0$. The substitution $t = \sqrt{(\alpha/p)}\mathrm{e}^s$ in the transform integral (1.1.1) gives (Watson, 1944, Section 6.22)

$$\begin{aligned}\bar{u}(p) &= \left(\frac{\alpha}{p}\right)^{(\nu+1)/2}\int_{-\infty}^{\infty}\exp[(\nu+1)s - 2\sqrt{(\alpha p)}\cosh s]\,\mathrm{d}s \\ &= 2\left(\frac{\alpha}{p}\right)^{(\nu+1)/2}\int_0^\infty \exp[-2\sqrt{(\alpha p)}\cosh s] \\ &\qquad \cosh[(\nu+1)s]\,\mathrm{d}s \qquad\qquad (1.3.1)\\ &= 2(\alpha/p)^{(\nu+1)/2}\mathrm{K}_{\nu+1}(2\sqrt{(\alpha p)}).\end{aligned}$$

**E1.3.10** In particular, since (Watson, 1944, Section 3.71)

$$\mathrm{K}_{1/2}(z) = \sqrt{\left(\frac{\pi}{2z}\right)}\mathrm{e}^{-z},$$

$$\mathscr{L}\{t^{-1/2}\exp(-\alpha/t)\} = \sqrt{(\pi/p)}\exp[-2\sqrt{(\alpha p)}].$$

This can also be obtained by the substitution $x = \sinh(\frac{1}{2}s)$ in the integral (1.3.1), which gives

$$\mathscr{L}\{t^{1/2}\exp(-\alpha/t)\} = 4(\alpha/p)^{1/4}\int_0^\infty \exp[-2\sqrt{(\alpha p)}(1 + 2x^2)]\mathrm{d}x,$$

from which the same result follows. Similarly, or by differentiation with respect to $\alpha$,

$$\mathscr{L}\{t^{-3/2}\exp(-\alpha/t)\} = \sqrt{(\pi/\alpha)}\exp[-2\sqrt{(\alpha p)}].$$

**E1.3.11** The results of E1.3.9 and E1.3.10 also apply when $\mathscr{R}\alpha = 0$, provided that $\mathscr{R}v > -1$. Hence, taking $\alpha = \pm i$, we obtain

$$\mathscr{L}\{t^{-1/2}\cos(t^{-1})\} = \sqrt{(\pi/p)}\exp[-\sqrt{(2p)}]\cos[\sqrt{(2p)}]$$
$$\mathscr{L}\{t^{-1/2}\sin(t^{-1})\} = \sqrt{(\pi/p)}\exp[-\sqrt{(2p)}]\sin[\sqrt{(2p)}].$$

Another useful method for finding the transform $\bar{u}(p)$ is to start from a differential equation satisfied by $u(t)$. Examples of this will be given in Section 2.4.

## 1.4 Elementary Methods of Inversion

Although the most powerful method of finding the inverse transform of a given function of $p$ is that of contour integration, described in Section 3.3, there are some useful methods based on the properties given in this chapter. In writing $u(t) = \mathscr{L}^{-1}\{\bar{u}(p)\}$, we assume that the function $u(t)$ is defined uniquely by its transform $\bar{u}(p)$. This will be justified in Section 3.1 on the hypothesis that $u(t)$ is continuous.

The partial-fraction method based on P1.1.8 has already been described. As seen in E1.1.9, this method is awkward when there are repeated quadratic factors, and it may be easier to apply the convolution property P1.2.2 or to employ differentiation with respect to $p$ or a parameter. It is also helpful to simplify the transform by using P1.1.3.

**E1.4.1** If

$$\bar{u}(p) = \frac{p+3}{(p^2+2p+5)^2} = \frac{(p+1)+2}{[(p+1)^2+4]^2},$$

then $u(t) = e^{-t}v(t)$, where

$$\bar{v}(p) = \frac{p+2}{(p^2+4)^2}.$$

**E1.4.2**

$$\begin{aligned}\mathscr{L}^{-1}\{p^{-3}(p+1)^{-2}\} &= (\tfrac{1}{2}t^2)*(te^{-t})\\ &= \int_0^t \tfrac{1}{2}s^2(t-s)e^{-(t-s)}\,ds\\ &= [(\tfrac{1}{2}s^2t - \tfrac{1}{2}s^3 - st + \tfrac{3}{2}s^2 + t - 3s + 3)e^{s-t}]_0^t\\ &= \tfrac{1}{2}t^2 - 2t + 3 - (t+3)e^{-t}\end{aligned}$$

as in E1.1.8. Here we have used P1.2.2 and equation (1.1.4).

**E1.4.3**

$$(p^2+\omega^2)^{-2} = \frac{1}{\omega^2}\,\frac{\omega}{p^2+\omega^2}\,\frac{\omega}{p^2+\omega^2}$$

Hence

$$\begin{aligned}\mathscr{L}^{-1}\{(p^2+\omega^2)^{-2}\} &= \omega^{-2}\sin\omega t * \sin\omega t\\ &= \omega^{-2}\int_0^t \sin\omega s\,\sin[\omega(t-s)]\,\mathrm{d}s\\ &= \tfrac{1}{2}\omega^{-3}(\sin\omega t - \omega t\cos\omega t).\end{aligned}$$

**E1.4.4** Similarly

$$\begin{aligned}\mathscr{L}^{-1}\{p(p^2+\omega^2)^{-2}\} &= \omega^{-1}\cos\omega t * \sin\omega t\\ &= \tfrac{1}{2}\omega^{-1}t\sin\omega t.\end{aligned}$$

This can also be obtained from E1.4.3 by using P1.1.4.

**E1.4.5**

$$(p^2+\omega^2)^{-2} = -\frac{1}{2\omega}\frac{\partial}{\partial\omega}\left(\frac{1}{p^2+\omega^2}\right).$$

Thus from P1.3.2

$$\begin{aligned}\mathscr{L}^{-1}\{(p^2+\omega^2)^{-2}\} &= -\frac{1}{2\omega}\frac{\partial}{\partial\omega}(\omega^{-1}\sin\omega t)\\ &= \tfrac{1}{2}\omega^{-3}(\sin\omega t - \omega t\cos\omega t).\end{aligned}$$

**E1.4.6**

$$p(p^2+\omega^2)^{-2} = -\frac{1}{2}\frac{\partial}{\partial p}\left(\frac{1}{p^2+\omega^2}\right).$$

From P1.1.9 we have

$$\mathscr{L}^{-1}\{p(p^2+\omega^2)^{-2}\} = \tfrac{1}{2}t(\omega^{-1}\sin\omega t).$$

**E1.4.7** Applying these results to E1.4.1, we have

$$\begin{aligned}v(t) &= \mathscr{L}^{-1}\left\{\frac{p+2}{(p^2+4)^2}\right\}\\ &= \tfrac{1}{4}t\sin 2t + \tfrac{1}{8}(\sin 2t - 2t\cos 2t),\end{aligned}$$

and so

$$u(t) = \mathscr{L}^{-1}\left\{\frac{p+3}{(p^2+2p+5)^2}\right\}$$

$$= \tfrac{1}{4}e^{-t}[(t+\tfrac{1}{2})\sin 2t - t\cos 2t],$$

as found in E1.1.9.

The method of series expansion (P1.3.1) can be applied to inverse transforms as follows.

**P1.4.1** If

$$\bar{u}(p) = \sum_{n=0}^{\infty} a_n p^{-n-\nu-1},$$

where $\mathscr{R}\nu > -1$ and the series converges for $|p| \geqslant R$, then

$$u(t) = \sum_{n=0}^{\infty} \frac{a_n}{\Gamma(n+\nu+1)} t^{n+\nu},$$

and this series converges for all $t$.

For any $A > R$,

$$\left|\frac{a_n}{\Gamma(n+\nu+1)}\right| \leqslant k\frac{A^n}{n!}.$$

See also P3.3.2.

The results of E1.4.3 and E1.4.4 can alternatively be obtained by this method.

**E1.4.8**

$$\bar{u}(p) = \sinh^{-1}[\sqrt{(\alpha/p)}] = \frac{1}{2}\log\left(\frac{\sqrt{(p+\alpha)}+\sqrt{\alpha}}{\sqrt{(p+\alpha)}-\sqrt{\alpha}}\right)$$

can be expanded for $|p| > |\alpha|$ as

$$\bar{u}(p) = \sum_{n=0}^{\infty} \frac{(-1)^n\frac{1}{2}\cdot\frac{3}{2}\ldots(n-\frac{1}{2})}{n!(2n+1)}\left(\frac{\alpha}{p}\right)^{n+1/2}$$

Hence its inverse transform is

$$u(t) = \sum_{n=0}^{\infty} \frac{(-1)^n\frac{1}{2}\cdot\frac{3}{2}\ldots(n-\frac{1}{2})}{n!(2n+1)}\,\frac{\alpha^{n+1/2}t^{n-1/2}}{\Gamma(n+\frac{1}{2})}$$

$$= \frac{1}{t\sqrt{\pi}}\sum_{n=0}^{\infty}\frac{(-1)^n(\alpha t)^{n+1/2}}{n!(2n+1)} = \frac{1}{2t}\operatorname{erf}(\sqrt{(\alpha t)}).$$

## 1.5 Additional Examples

**E1.5.1** Find the Laplace transforms of the functions
(i) $t(t+1)(t+2)$,
(ii) $(t+1)e^{-t}$,
(iii) $t^2 \cos t$,
(iv) $\sin^3 t$,
(v) $(t^2+1)\sinh 2t - 2t\cosh 2t$,
(vi) $\sinh t \sin t$.

**E1.5.2** What functions have the following Laplace transforms?
(i) $1/[p(p+1)(p+2)]$,
(ii) $(p-2)/(p^2+4p+3)$,
(iii) $(p+3)/(p^2+4p+5)$,
(iv) $(p^2-1)/[p(p^2+1)]$,
(v) $(p+1)/[p^2(p-1)]$,
(vi) $(p+1)/[p(p-1)^2]$.

**E1.5.3** Find the Laplace transform of $\cosh \alpha t \cos \alpha t$, where $\alpha$ is a constant, and verify that it satisfies P1.1.2. Use P1.1.4 to find the function whose transform is $\alpha^3/(p^4+4\alpha^4)$.

**D1.5.1** The *Laguerre polynomial* $L_n(t)$ is defined by

$$L_n(t) = \frac{e^t}{n!}\frac{d^n}{dt^n}(t^n e^{-t}) \qquad \text{for } n = 0, 1, 2, \ldots .$$

**E1.5.4** Show that

$$\mathscr{L}\{L_n(t)\} = \frac{(p-1)^n}{p^{n+1}},$$

and hence obtain the relations

$$L_n'(t) = L_{n-1}'(t) - L_{n-1}(t),$$
$$tL_n'(t) = n[L_n(t) - L_{n-1}(t)].$$

**E1.5.5** Prove by induction that

$$\mathscr{L}\{(1-e^{-\alpha t})^n\} = \frac{n!\,\alpha^n}{p(p+\alpha)\ldots(p+n\alpha)} \qquad \text{for } n = 0, 1, 2, \ldots .$$

Hence evaluate $\mathscr{L}\{\sin^n t\}$.

**E1.5.6** Find the Laplace transform of the function

$$u(t) = \begin{cases} t & (0 < t < a), \\ ae^{-(t-a)} & (t > a). \end{cases}$$

**E1.5.7** Show that the 'staircase' function, defined by

$$u(t) = na \qquad \text{for } na < t < (n+1)a, \quad n = 0, 1, 2, \ldots,$$

satisfies the equation

$$u(t) - u(t-a)\mathrm{H}(t-a) = a\mathrm{H}(t-a),$$

and hence find the transform $\bar{u}(p)$.

**E1.5.8** Given that

$$v(t) = \begin{cases} \sin t & (0 < t < \pi), \\ 0 & (t > \pi), \end{cases}$$

show that

$$v(t) = \sin t + \sin(t-\pi)\mathrm{H}(t-\pi)$$

and hence find its transform $\bar{v}(p)$. Deduce the transform of the rectified sine wave $u(t) = |\sin t|$.

**E1.5.9** Expresses

$$v(t) = \begin{cases} t & (0 < t < a), \\ 0 & (t > a) \end{cases}$$

in a form involving the unit function, and hence find $\bar{v}(p)$. Deduce the transform of the 'sawtooth' function given by

$$u(t) = t - na \qquad \text{for } na < t < (n+1)a, \quad n = 0, 1, 2, \ldots .$$

Check the result with the aid of E1.5.7.

**E1.5.10** Verify that $(e^{\alpha t} * e^{\beta t}) * e^{\gamma t} = e^{\alpha t} * (e^{\beta t} * e^{\gamma t})$ on the assumption that $\alpha$, $\beta$ and $\gamma$ are distinct.

**E1.5.11** Evaluate the convolution $\mathrm{H}(t-a) * \mathrm{H}(t-b)$, where $a$ and $b$ are positive, and check the result by finding its Laplace transform.

**E1.5.12** Show that if $w(t) = u(t) * v(t)$, then
(i) $[u(t)e^{\alpha t}] * [v(t)e^{\alpha t}] = w(t)e^{\alpha t}$,
(ii) $[u(t-a)\mathrm{H}(t-a)] * [v(t-b)\mathrm{H}(t-b)] = w(t-a-b)\mathrm{H}(t-a-b)$.

**E1.5.13** Let $I_{\mu,\nu}(t) = t^\mu * t^\nu$. Prove the recurrence relations

$$I_{\mu,\nu}(t) = \frac{\nu}{\mu+1} I_{\mu+1,\nu-1}(t) = \frac{\nu t}{\mu+\nu+1} I_{\mu,\nu-1}(t),$$

where $\mathscr{R}\mu > -1, \mathscr{R}\nu > 0$.

**E1.5.14** Show that, provided $a$ is positive or complex,

$$\mathscr{L}\{(t+a)^{-1}\} \sim \sum_{n=0}^{\infty} \frac{(-1)^n n!}{(ap)^{n+1}}$$

as $|p| \to \infty$ in $|\arg p| \leqslant \frac{1}{2}\pi - \delta$.

**E1.5.15** Show that if $\mathscr{R}\nu > 0, \mathscr{R}\alpha > 0$, then

$$\mathscr{L}\{(1-e^{-\alpha t})^{\nu-1}\} = \frac{\Gamma(\nu)\Gamma(p/\alpha)}{\alpha\Gamma(\nu+p/\alpha)},$$

and compare with E1.5.5. From the case $\alpha = 1$, prove that

$$\frac{\Gamma(p+\nu)}{\Gamma(p)} \sim p^\nu$$

as $|p| \to \infty$ with $|\arg p| \frac{1}{2}\pi - \delta$, and show that this result holds throughout $|\arg p| \leqslant \pi \leqslant \delta$ by considering $\alpha$ to be complex.

**E1.5.16** Obtain the transforms of
(i) $t^{-1} \sin t$,
(ii) $t^{-1}(1 - e^{-\alpha t})$,

by expanding in powers of $t$.

**E1.5.17** Show that if $\mathscr{R}\nu > -\frac{1}{2}$

$$\mathscr{L}\{t^\nu I_\nu(\alpha t)\} = \frac{\Gamma(2\nu+1)}{\Gamma(\nu+1)} (\tfrac{1}{2}\alpha)^\nu (p^2 - \alpha^2)^{-\nu-1/2}.$$

[The result may be expressed by means of the duplication formula (E1.5.38) as $\pi^{-1/2}\Gamma(\nu+\frac{1}{2})(2\alpha)^\nu(p^2-\alpha^2)^{-\nu-1/2}$.]

**D1.5.2** The *hypergeometric function* $F(a, b; c; z)$ is defined for $|z| < 1$ as

$$F(a, b; c; z) = \sum_{n=0}^{\infty} \frac{(a)_n (b)_n}{(c)_n n!} z^n,$$

where

$$(a)_n = a(a+1)\dots(a+n-1) = \frac{\Gamma(a+n)}{\Gamma(a)}.$$

The general solution of the *hypergeometric equation*

$$z(1-z)\frac{d^2w}{dz^2} + [c-(a+b+1)z]\frac{dw}{dz} - abw = 0$$

is

$$w = AF(a,b;c;z) + Bz^{1-c}F(a-c+1, b-c+1; 2-c; z),$$

provided that $c$ is not an integer.

**D1.5.3** The *confluent hypergeometric function* ${}_1F_1(a;c;z)$ is defined for all $z$ by

$${}_1F_1(a;c;z) = \sum_{n=0}^{\infty} \frac{(a)_n z^n}{(c)_n n!}.$$

The general solution of the *confluent hypergeometric equation*

$$z\frac{d^2w}{dz^2} + (c-z)\frac{dw}{dz} - aw = 0$$

is

$$w = A\,{}_1F_1(a;c;z) + Bz^{1-c}\,{}_1F_1(a-c+1;2-c;z),$$

provided that $c$ is not an integer. The particular solution $U(a;c;z)$ is obtained by taking

$$A = \frac{\Gamma(1-c)}{\Gamma(a-c+1)}, \qquad B = \frac{\Gamma(c-1)}{\Gamma(a)},$$

or as the limit of this expression. Further properties are given by Slater (1960).

**E1.5.18** Show that if $\mathscr{R}b > 0$, then
(i) $\mathscr{L}\{t^{b-1}\,{}_1F_1(a;c;\alpha t)\} = \Gamma(b)p^{-b}F(a,b;c;\alpha/p)$,
and in particular that if $\mathscr{R}c > 0$
(ii) $\mathscr{L}\{t^{c-1}\,{}_1F_1(a;c;\alpha t)\} = \Gamma(c)p^{a-c}(p-\alpha)^{-a}$.

**E1.5.19** From E1.5.18 (ii), derive the formulae
(i) ${}_1F_1(a;c;-t) = e^{-t}\,{}_1F_1(c-a;c;t)$,
(ii) ${}_1F_1(a;2a;t) = \Gamma(a+\frac{1}{2})(\frac{1}{4}t)^{1/2-a}e^{t/2}\mathrm{I}_{a-1/2}(\frac{1}{2}t)$.
[These are *Kummer's first* and *second theorems.*]

**E1.5.20** Show that

$$\mathscr{L}\{t^{\mu-1}(1-t)^{\nu-1}\mathrm{H}(1-t)\} = \frac{\Gamma(\mu)\Gamma^{s}(\nu)}{\Gamma(\mu+\nu)}\,{}_1F_1(\mu;\mu+\nu;-p)$$

for $\mathscr{R}\mu > 0, \mathscr{R}\nu > 0$. Hence obtain the asymptotic expansion of ${}_1F_1(a;c;-p)$ as $|p| \to \infty$ with $|\arg p| \leqslant \frac{1}{2}\pi - \delta$.

**E1.5.21** Show that if $\mathscr{R}\nu > -1$

$$\mathscr{L}\{t^{-1/2}\mathrm{I}_\nu(2\sqrt{(\alpha t)})\} = \sqrt{\left(\frac{\pi}{p}\right)}\exp\left(\frac{\alpha}{2p}\right)\mathrm{I}_{\nu/2}\left(\frac{\alpha}{2p}\right).$$

[E1.5.19 (ii) has been used, and the result simplified with the aid of the duplication formula (E1.5.38).]

**E1.5.22** From E1.5.15, show by differentiation that

$$\mathscr{L}\{\log(1-\mathrm{e}^{-\alpha t})\} = -\left[\psi\left(\frac{p}{\alpha}+1\right)+\gamma\right]p^{-1},$$

where $\mathscr{R}\alpha > 0$. Deduce that

(i) $\mathscr{L}\{t^{-1}-\alpha(1-\mathrm{e}^{-\alpha t})^{-1}\} = \psi\left(\frac{p}{\alpha}\right) - \log\left(\frac{p}{\alpha}\right),$

(ii) $\mathscr{L}\{t(1-\mathrm{e}^{-\alpha t})^{-1}\} = \frac{1}{\alpha^2}\psi'\left(\frac{p}{\alpha}\right).$

**E1.5.23** Evaluate $\mathscr{L}\{t^{-2}(1-\cos\alpha t)\}$.

**E1.5.24** Find the transforms of $E_1(t)$ and $\mathrm{Si}(t)$ (D1.3.4 and D1.3.5).

**E1.5.25** Use the convolution property P1.2.2 to find the inverse transforms of

(i) $\dfrac{1}{(p+1)^2(p-1)},$

(ii) $\dfrac{1}{p(p^2-1)^2},$

(iii) $\dfrac{p-1}{(p+1)(p^2+1)^2},$

(iv) $\dfrac{20p}{(p^2+4p+8)(p^2+8p+20)}.$

**E1.5.26** Show that $\mathscr{L}^{-1}\{p^{-1}(p+\alpha)^{-1/2}\} = \alpha^{-1/2}\,\text{erf}(\sqrt{(\alpha t)})$. Hence obtain the inverse transforms of
(i) $p^{-1}(p+\alpha)^{1/2}$,
(ii) $(p^{1/2}+\alpha^{1/2})^{-1}$.

**E1.5.27** Find the inverse transform of

$$p\log\left(\frac{p^2+\alpha^2}{p^2-\alpha^2}\right)$$

(*a*) by differentiation, (*b*) by series expansion.

**E1.5.28** Evaluate the Laplace transform of

$$\text{erfc}\left(\frac{\alpha}{\sqrt{t}}+\beta\sqrt{t}\right),$$

where $|\arg\alpha| < \frac{1}{4}\pi$. Hence find the inverse transform of

$$p^{-1}\exp[-2\alpha\sqrt{(p+\beta^2)}].$$

**E1.5.29** If

$$\bar{u}(p;\alpha,\beta) = \frac{\exp(-\alpha\sqrt{p})}{p+\beta\sqrt{p}},$$

where $|\arg\alpha| < \frac{1}{4}\pi$, show that

$$\frac{\partial u}{\partial\alpha} - \beta u = -\frac{1}{\sqrt{(\pi t)}}\exp\left(-\frac{\alpha^2}{4t}\right),$$

and deduce that

$$u(t;\alpha,\beta) = \exp(\alpha\beta+\beta^2 t)\,\text{erfc}\left(\frac{\alpha}{2\sqrt{t}}+\beta\sqrt{t}\right).$$

**E1.5.30** The Laplace transform $\mathscr{L}\{u(t)\}$ converges *simply* (not necessarily absolutely) for $p=\alpha$, and

$$v(t) = \int_0^t u(s)\mathrm{e}^{-\alpha s}\,\mathrm{d}s.$$

Prove that
(i) $\mathscr{L}\{v(t)\}$ converges absolutely if $\mathscr{R}p > 0$;
(ii) if $\mathscr{R}(p-\alpha) > 0$ then $\mathscr{L}\{u(t)\}$ converges simply to

$$\bar{u}(p) = (p-\alpha)\bar{v}(p-\alpha);$$

(iii) $\mathscr{L}\{u(t)\}$ converges uniformly in the sector $S$ given by

$$|\arg(p-\alpha)| \leqslant \tfrac{1}{2}\pi - \delta;$$

(iv) $\bar{u}(p) \to \bar{u}(\alpha)$ as $p \to \alpha$ in $S$.

**E1.5.31** Use E1.5.30 to prove that

$$\int_0^\infty t^{-1} \sin t \, \mathrm{d}t = \tfrac{1}{2}\pi.$$

**E1.5.32** Show that $\mathscr{L}\{t\cos(\exp t^2)\exp(t^2/2)\}$ converges simply for all $p$, but absolutely for no $p$.

**E1.5.33** (i) Prove that, provided a certain double integral converges absolutely,

$$\mathscr{L}\{\bar{u}(t)v(t)\} = \int_0^\infty u(s)\bar{v}(p+s)\,\mathrm{d}s.$$

(ii) Show that P1.3.3 may be generalized to

$$\mathscr{L}\{t^{-\nu}v(t)\} = \frac{1}{\Gamma(\nu)}\int_p^\infty (z-p)^{\nu-1}\bar{v}(z)\,\mathrm{d}z,$$

where $\mathscr{R}\nu > 0$.

(iii) Show that if $\bar{u}(p)$ converges for all real $p > 0$ and $\bar{u}(t)$ has a Laplace transform, it is given by

$$\bar{\bar{u}}(p) = \int_0^\infty \frac{u(s)}{p+s}\,\mathrm{d}s \quad \text{for } \mathscr{R}p > 0.$$

[This is the *Stieltjes transform* of $u(t)$.]

**E1.5.34** The function $u(t)$ is differentiable for $t > 0$ except at $t = t_1, t_2, t_3, \ldots$, where it has simple discontinuities. Show that if $u'(t)$ has a Laplace transform, it is given by

$$\mathscr{L}\{u'(t)\} = p\bar{u}(p) - u(0) - \sum_n [u(t_n)]\exp(-pt_n)$$

where $[u(t_n)] = u(t_n+) - u(t_n-)$.

**D1.5.4** *Dawson's function* is defined by

$$F(t) = \exp(-t^2)\int_0^t \exp(s^2)\,\mathrm{d}s.$$

**E1.5.35** Show that the Laplace transform of Dawson's function, is

$$\bar{F}(p) = 4\exp(\tfrac{1}{4}p^2)E_1(\tfrac{1}{4}p^2).$$

**D1.5.5** *Scorer's function* is defined by

$$\mathrm{Hi}(-p) = \mathscr{L}\{\pi^{-1}\exp(-\tfrac{1}{3}t^3)\}.$$

**E1.5.36** The function $x(t)$ is the real root of the equation $x^3 + ax - t = 0$, where $a > 0$. Show that

$$\bar{x}(p) = 3^{-1/3}\pi p^{-4/3}\mathrm{Hi}(-3^{-1/3}ap^{2/3}).$$

Discuss the case $a < 0$.

**E1.5.37** Let $u(t) = \mathrm{J}_\nu(2\sqrt{(\alpha t)})\mathrm{J}_\nu(2\sqrt{(\beta t)})$, where $\mathscr{R}\nu > -1$. Show that

$$\bar{u}(p) = \sum_{m=0}^{\infty} \frac{(-\alpha)^m(\alpha\beta)^{\nu/2}}{m!\,\Gamma(\nu+1)p^{m+\nu+1}}\ {}_1F_1\left(m+\nu+1;\ \nu+1;\ -\frac{\beta}{p}\right),$$

and use Kummer's first theorem (E1.5.19 (i)) to obtain

$$\bar{u}(p) = \frac{(\alpha\beta)^{\nu/2}}{\Gamma(\nu+1)}\exp\left(-\frac{\beta}{p}\right)\sum_{m=0}^{\infty}\frac{(-\alpha)^m}{p^{m+\nu+1}}\sum_{n=0}^{m}\frac{(-\beta/p)^n}{(\nu+1)_n(m-n)!\,n!}.$$

By reversing the order of summation, show that

$$\bar{u}(p) = p^{-1}\exp\left(-\frac{\alpha+\beta}{p}\right)\mathrm{I}_\nu\left(2\frac{\sqrt{(\alpha\beta)}}{p}\right).$$

**D1.5.6** The *Bernoulli numbers* $B_n$ are given by

$$\frac{z}{1-\mathrm{e}^{-z}} = 1 + \tfrac{1}{2}z + \sum_{n=1}^{\infty}\frac{(-1)^{n-1}B_n}{(2n)!}z^{2n} \qquad \text{for } |z| < 2\pi.$$

**E1.5.38** From E1.5.22 (i), prove that

(i) $\psi(p) \sim \log p - \tfrac{1}{2}p^{-1} + \sum_{n=1}^{\infty}\frac{(-1)^n B_n}{2n}p^{-2n}$

as $|p| \to \infty$ with $|\arg p| \leqslant \pi - \delta$;
(ii) $2\psi(2p) = \psi(p) + \psi(p+\tfrac{1}{2}) + 2\log 2$.
Deduce from (ii) the *duplication formula* for the gamma function, namely

$$\Gamma(\tfrac{1}{2})\Gamma(2p) = 2^{2p-1}\Gamma(p)\Gamma(p+\tfrac{1}{2}).$$

**D1.5.7** Convolution powers of a function may be defined by

$$u(t)^{*1} = u(t), \qquad u(t)^{*(n+1)} = u(t) * [u(t)^{*n}],$$

in the case of positive integer powers, and more generally by

$$u(t)^{*\nu} = \mathscr{L}^{-1}\{(\bar{u}(p))^{\nu}\} \qquad \text{for } \nu > 0$$

(and possibly for some complex values of $\nu$).

**E1.5.39** Show that

(i) $1^{*\nu} = \dfrac{t^{\nu-1}}{\Gamma(\nu)}$;

(ii) $(\sin t)^{*1/2} = \mathrm{J}_0(t)$;

(iii) $(1 - \mathrm{e}^{-t})^{*\nu} = \dfrac{\Gamma(\nu + \frac{1}{2})}{\Gamma(2\nu)}(4t)^{\nu - 1/2}\mathrm{e}^{-t/2}\mathrm{I}_{\nu - 1/2}(\frac{1}{2}t)$.

**E1.5.40** If $v(t) = u(t) * (1^{*\nu})$, where $0 < \nu < 1$, show that

$$u(t) = \frac{\mathrm{d}}{\mathrm{d}t}[v(t) * (1^{*(1-\nu)})].$$

**E1.5.41** Prove that P1.2.13 may be generalized, under the same conditions, as follows. If

$$u(t) = \sum_{n=0}^{\infty} (a_n + b_n \log t)t^{n\mu + \nu - 1} \qquad \text{as } t \to 0+,$$

then

$$\bar{u}(p) \sim \sum_{n=0}^{\infty} \{a_n + b_n[\psi(n\mu + \nu) - \log p]\} \frac{\Gamma(n\mu + \nu)}{p^{n\mu+\nu}}$$

as $|p| \to \infty$ with $|\arg p| \leqslant \frac{1}{2}\pi - \delta$.

**D1.5.8** The *double Laplace transform* of a function $u(x, y)$ defined in $x > 0$, $y > 0$ is

$$\bar{u}(p, q) = \int_0^{\infty}\int_0^{\infty} u(x, y)\mathrm{e}^{-px-qy}\,\mathrm{d}x\,\mathrm{d}y.$$

**E1.5.42** Show that

(i) if $u(x, y) = v(x + y)$ then

$$\bar{u}(p, q) = \frac{\bar{v}(p) - \bar{v}(q)}{q - p};$$

(ii) if

$$u(x, y) = \begin{cases} v(x-y) & (x > y), \\ w(y-x) & (y > x), \end{cases}$$

then

$$\bar{u}(p, q) = \frac{\bar{v}(p) + \bar{w}(q)}{p+q};$$

(iii) if $u(x, y) = v(xy)$ then

$$\bar{u}(p, q) = \int_0^\infty v(z) \mathrm{K}_0(2\sqrt{(pqz)}) \mathrm{d}z;$$

(iv) if $u(x, y) = v(y/x)$ then

$$\bar{u}(p, q) = \int_0^\infty \frac{v(z)}{(p+qz)^2} \mathrm{d}z.$$

# 2
# Ordinary Differential Equations

## 2.1 Equations with Constant Coefficients

Laplace transforms are particularly well suited to the solution of ordinary differential equations when these are linear with constant coefficients, when the forcing term is an exponential-type function, and when the solution is made definite by initial conditions. The effect of applying the Laplace transformation is then to replace the differential equation by a linear algebraic equation for the transformed function, as in the following example.

**E2.1.1**

$$\frac{d^4u}{dt^4} - 2\frac{d^2u}{dt^2} + u = e^t,$$

with initial conditions $u(0) = 1$, $u'(0) = u''(0) = 0$, $u'''(0) = -1$.

The Laplace transform of the equation is, from P1.1.5,

$$p^4\bar{u} - p^3 + 1 - 2(p^2\bar{u} - p) + \bar{u} = (p-1)^{-1},$$

so that

$$(p^2-1)^2\bar{u}(p) = (p-1)^{-1} + p^3 - 2p - 1.$$

Hence

$$\bar{u}(p) = \frac{1}{(p-1)^3(p+1)^2} + \frac{p^2-p-1}{(p-1)^2(p+1)}$$
$$= \frac{1}{4(p-1)^3} - \frac{3}{4(p-1)^2} + \frac{15}{16(p-1)} - \frac{1}{8(p+1)^2}$$
$$+ \frac{1}{16(p+1)},$$

and this is the transform of

$$u(t) = (\tfrac{1}{8}t^2 - \tfrac{3}{4}t + \tfrac{15}{16})e^t - (\tfrac{1}{8}t - \tfrac{1}{16})e^{-t}.$$

It is a good practice to check answers by substitution into the given differential equation and boundary conditions.

The general equation of the type considered here may be written as

$$a_n \frac{\mathrm{d}^n u}{\mathrm{d}t^n} + a_{n-1} \frac{\mathrm{d}^{n-1} u}{\mathrm{d}t^{n-1}} + \dots + a_0 u = g(t), \tag{2.1.1a}$$

where $a_0, a_1, \dots, a_n$ are constants with $a_n \neq 0$, or as

$$P(\mathrm{D})u(t) = g(t), \tag{2.1.1b}$$

where $\mathrm{D} \equiv \mathrm{d}/\mathrm{d}t$ and $P(\mathrm{D})$ is the polynomial

$$P(\mathrm{D}) = \sum_{r=0}^{n} a_r \mathrm{D}^r.$$

In equations (2.1.1), $u(t)$ is the unknown function and $g(t)$, the forcing term, is a given function. Since (2.1.1) is an $n$th-order equation, we expect to be able to impose $n$ initial conditions

$$\mathrm{D}^r u(0) = c_r \qquad (r = 0, 1, \dots, n-1). \tag{2.1.2}$$

We can justify the operation of taking the Laplace transform of equation (2.1.1) by the following result.

**P2.1.1** If $g(t)$ is an exponential-type function, then any solution of (2.1.1) is also an exponential-type function.

Since the operator D commutes with multiplication by a constant, we can factorize the polynomial $P(\mathrm{D})$ and write (2.1.1) in the form

$$a_n(\mathrm{D} - \alpha_1)(\mathrm{D} - \alpha_2)\dots(\mathrm{D} - \alpha_n)u(t) = g(t),$$

where $\alpha_1, \alpha_2, \dots, \alpha_n$ are the roots (repeated according to multiplicity) of the algebraic equation $P(\alpha) = 0$. Equation (2.1.1) is therefore equivalent to the system of equations

$$(\mathrm{D} - \alpha_m)u_m(t) = u_{m-1}(t) \qquad (m = 1, 2, \dots, n),$$

where $u_0(t) = g(t)/a_n$ and $u(t) = u_n(t)$.

When $u_{m-1}(t)$ is known

$$u_m(t) = \exp(\alpha_m t)\left(\int_0^t \exp(-\alpha_m s)\, u_{m-1}(s)\mathrm{d}s + k_m\right),$$

where $k_m$ is a constant of integration. By hypothesis, $u_0(t)$ is an e.t.f. and, from P1.1.7, if $u_{m-1}(t)$ is an e.t.f., so also is $u_m(t)$. Thus, by induction, each $u_m(t)$ is an e.t.f. and in particular $u(t)$ is.

To illustrate P2.1.1, consider the equation of E2.1.1.

**E2.1.2** $(D^4 - 2D^2 + 1)u(t) = e^t$.
Since $D^4 - 2D^2 + 1 = (D-1)^2(D+1)^2$, the equation is equivalent to the system

$$\begin{aligned} Du_1 - u_1 &= e^t, \\ Du_2 - u_2 &= u_1, \\ Du_3 + u_3 &= u_2, \\ Du + u &= u_3. \end{aligned}$$

Solving these in turn gives us

$$\begin{aligned} e^{-t}u_1(t) &= t + k_1, \\ e^{-t}u_2(t) &= \tfrac{1}{2}t^2 + k_1 t + k_2, \\ e^t u_3(t) &= [\tfrac{1}{4}t^2 + (\tfrac{1}{2}k_1 - \tfrac{1}{4})t + \tfrac{1}{2}k_2 - \tfrac{1}{4}k_1 + \tfrac{1}{8}]e^{2t} + k_3, \\ u(t) &= [\tfrac{1}{8}t^2 + (\tfrac{1}{4}k_1 - \tfrac{1}{4})t + \tfrac{1}{4}k_2 - \tfrac{1}{4}k_1 + \tfrac{3}{16}]e^t + (k_3 t + k_4)e^{-t}. \end{aligned}$$

The constants $k_1, k_2, k_3$ and $k_4$ are those that appear naturally when this system is solved, but a different ordering of the factors of the differential operator would give a different system, and therefore different constants in $u(t)$. The solution can be expressed more simply as

$$u(t) = (\tfrac{1}{8}t^2 + k_1' t + k_2')e^t + (k_3 t + k_4)e^{-t}.$$

Here $\frac{1}{8}t^2 e^t$ is a *particular integral* of the original equation, and the arbitrary constants $k_1', k_2', k_3$ and $k_4$ multiply four independent *complementary functions* $te^t, e^t, te^{-t}$ and $e^{-t}$ that satisfy the corresponding homogeneous equation $(D^4 - 2D^2 + 1)u = 0$.

The method of P2.1.1 shows that the general solution $u(t)$ of (2.1.1) has $n$ arbitrary constants $k_1, k_2, \ldots, k_n$. The particular solution that satisfies the initial conditions (2.1.2) may be obtained by the Laplace transform method as follows.

**P2.1.2** If $u(t)$ is the solution of the differential equation

$$P(D)u(t) \equiv \sum_{r=0}^{n} a_r D^r u(t) = g(t) \tag{2.1.1}$$

that satisfies the initial conditions

$$D^r u(0) = c_r \qquad (r = 0, 1, \ldots, n-1), \tag{2.1.2}$$

then its Laplace transform is given by

$$\bar{u}(p) = \frac{Q(p)}{P(p)},$$

where

$$Q(p) = \bar{g}(p) + \sum_{s=0}^{n-1} c_s \sum_{r=s+1}^{n} a_r p^{r-s-1}.$$

From P1.1.5, the Laplace transform of equation (2.1.1) is

$$\sum_{r=0}^{n} a_r [p^r \bar{u}(p) - c_0 p^{r-1} - c_1 p^{r-2} - \dots - c_{r-1}] = \bar{g}(p),$$

since P2.1.1 shows that each term of the equation has a transform.

Since $g(t)$ is an e.t.f., $\bar{g}(p)$ is a rational function of $p$ vanishing as $|p| \to \infty$. The rest of $Q(p)$ is a polynomial of degree $n-1$ at most. As $P(p)$ is a polynomial of degree $n$, P2.1.2 shows that $\bar{u}(p)$ is a rational function of $p$, and $\bar{u}(p) \to 0$ as $|p| \to \infty$. Thus from P1.1.8 $\bar{u}(p)$ is the transform of an e.t.f. $u(t)$.

Although the forcing term $g(t)$ has so far been taken to be an e.t.f., the Laplace transform method can be applied to other cases if we assume that the relevant transforms exist. An important class of forcing functions consists of those that act for only a limited interval of $t$, or that change their analytical forms at certain values of $t$. The Laplace transformation can then be applied separately to the initial and later forms of the differential equation, using the fact that in this equation all the derivatives except the highest must remain continuous. Their values at the point of change then provide initial conditions for the next interval. A more direct procedure is to use the delay operator of P1.1.10. Both methods are illustrated in the following example.

**E2.1.3**

$$(\mathrm{D}^2 + 1)u(t) = \begin{cases} 1 & (0 < t < T), \\ 0 & (t > T), \end{cases}$$

with $u(0) = u'(0) = 0$.

($a$) First consider the problem

$$\frac{\mathrm{d}^2 u}{\mathrm{d}t^2} + u = 1 \qquad \text{for all } t > 0$$

with $u(0) = u'(0) = 0$. Then $p^2\bar{u} + \bar{u} = p^{-1}$, so that

$$\bar{u}(p) = \frac{1}{p(p^2+1)} = \frac{1}{p} - \frac{p}{p^2+1}.$$

Thus $u(t) = 1 - \cos t$, and hence $u(T) = 1 - \cos T, u'(T) = \sin T$.

Now let $t = T + s, u(t) = v(s)$ and consider

$$\frac{d^2v}{ds^2} + v = 0 \qquad \text{for } s > 0,$$

with $v(0) = 1 - \cos T, v'(0) = \sin T$. Let $q$ be the transform variable corresponding to $s$. Then

$$q^2\bar{v} - q(1 - \cos T) - \sin T + \bar{v} = 0,$$

from which

$$\bar{v}(q) = \frac{(1 - \cos T)q + \sin T}{q^2 + 1},$$

and so

$$v(s) = (1 - \cos T)\cos s + \sin T \sin s.$$

In terms of the original variables, this is

$$u(t) = \cos(t - T) - \cos t \qquad \text{for } t > T.$$

Note that we have used the principle that the solution for $u(t)$ in $0 < t < T$ is independent of the form of the equation for $t > T$.

(*b*) Write the equation as

$$\frac{d^2u}{dt^2} + u = H(t) - H(t - T)$$

with $u(0) = u'(0) = 0$. Then, from P1.1.10, $(p^2 + 1)\bar{u}(p) = p^{-1}(1 - e^{-pT})$ and so

$$\bar{u}(p) = \left(\frac{1}{p} - \frac{p}{p^2 + 1}\right)(1 - e^{-pT}).$$

Consequently

$$u(t) = (1 - \cos t)H(t) - [1 - \cos(t - T)]H(t - T)$$
$$= \begin{cases} 1 - \cos t & (0 < t < T), \\ \cos(t - T) - \cos t & (t > T). \end{cases}$$

The result of P2.1.2 leads to

$$u(t) = v(t) + \sum_{s=0}^{n-1} c_s w_s(t),$$

where

$$\bar{v}(p) = \frac{\bar{g}(p)}{P(p)}, \qquad \bar{w}_s(p) = \frac{1}{P(p)} \sum_{r=s+1}^{n} a_r p^{r-s-1}.$$

Then $v(t)$ is the particular integral of (2.1.1) such that $\mathrm{D}^r v(0) = 0$ for $r = 0, 1, 2, \ldots, n-1$. Each $w_s(t)$ is a complementary function, so that $P(\mathrm{D})w_s(t) = 0$, and is defined by the initial conditions

$$D^r w_s(0) = \delta_{rs} \qquad \text{for } r = 0, 1, \ldots, n-1,$$

where

$$\delta_{rs} = \begin{cases} 0 & \text{if} \quad r \neq s, \\ 1 & \text{if} \quad r = s. \end{cases}$$

Each $w_s(t)$ is an e.t.f., and if $g(t)$ is an e.t.f. so also is $v(t)$.

The convolution property P1.2.2 shows that

$$v(t) = g(t) * w_{n-1}(t)/a_n,$$

and the form of $\bar{w}_s(p)$ shows that

$$w_s(t) = \sum_{r=s+1}^{n} a_r \mathrm{D}^{r-s-1} w_{n-1}(t)/a_n.$$

The function $w_{n-1}(t)/a_n$ represents the response of a system governed by equation (2.1.1) to a unit impulse at $t = 0$, that is to the case $g(t) = \delta(t)$.

The Laplace transform method is less useful when the conditions on $u(t)$ are not all at $t = 0$, but it can sometimes be applied by leaving one or more of the initial values as unknown.

**E2.1.4** $(\mathrm{D}^2 - 1)^2 u(t) = \mathrm{e}^{-t}$, with $u(0) = 1, u'(0) = 0$ and $u(t) \to 0$ as $t \to \infty$. It will appear that the condition at infinity is a double condition. Suppose that $u''(0) = a$ and $u'''(0) = b$. Then

$$p^4 \bar{u} - p^3 - ap - b - 2(p^2 \bar{u} - p) + \bar{u} = \frac{1}{p+1},$$

so that

$$(p^2 - 1)^2 \bar{u}(p) = \frac{1}{p+1} + p^3 + (a-2)p + b \equiv f(p), \qquad \text{say.}$$

When $\bar{u}(p)$ is expressed in partial fractions, it has the form

$$\bar{u}(p) = \frac{A}{(p+1)^3} + \frac{B}{(p+1)^2} + \frac{C}{p+1} + \frac{K}{(p-1)^2} + \frac{L}{p-1},$$

which corresponds to

$$u(t) = (\tfrac{1}{2}At^2 + Bt + C)\mathrm{e}^{-t} + (Kt + L)\mathrm{e}^t.$$

The condition that $u(t) \to 0$ as $t \to \infty$ is that $K = L = 0$, and hence that $\bar{u}(p)$ remains finite at $p = 1$. This requires that $f(1) = f'(1) = 0$, that is $a + b = \frac{1}{2}$ and $a = -\frac{3}{4}$. Hence $b = \frac{5}{4}$ and we obtain

$$\bar{u}(p) = \frac{p^2 + 3p + \frac{9}{4}}{(p+1)^3} = \frac{1}{4(p+1)^3} + \frac{1}{(p+1)^2} + \frac{1}{p+1},$$

so that

$$u(t) = (\tfrac{1}{8}t^2 + t + 1)e^{-t}.$$

The following example will be useful in Chapter 4.

**E2.1.5** $(D^2 - k^2)u(t) = g(t)$ for $0 < t < l$, with $u(0) = 1, u(l) = 0$.

We can define $g(t)$ arbitrarily for $t > l$, so long as $\bar{g}(p)$ exists. Then if $u'(0) = a$ the transform of the equation is

$$p^2\bar{u} - p - a - k^2\bar{u} = \bar{g}(p),$$

so that

$$\bar{u}(p) = \frac{p+a}{p^2 - k^2} + \frac{\bar{g}(p)}{p^2 - k^2}.$$

From P1.2.2

$$u(t) = \cosh kt + \frac{a}{k}\sinh kt + \frac{1}{k}g(t) * \sinh kt.$$

The condition $u(l) = 0$ therefore gives

$$\cosh kl + \frac{a}{k}\sinh kl + \frac{1}{k}\int_0^l g(s)\sinh[k(l-s)]\,ds = 0,$$

so that

$$u'(0) = a = -k\coth kl - \int_0^l g(s)\frac{\sinh[k(l-s)]}{\sinh kl}\,ds.$$

Hence

$$\begin{aligned} u(t) &= \cosh kt + \frac{1}{k}\int_0^t g(s)\sinh[k(t-s)]\,ds \\ &\quad - \frac{\sinh kt}{\sinh kl}\left(\cosh kl + \frac{1}{k}\int_0^l g(s)\sinh[k(l-s)]\,ds\right) \\ &= \frac{\sinh[k(l-t)]}{\sinh kl}\left(1 - \frac{1}{k}\int_0^t g(s)\sinh ks\,ds\right) \\ &\quad - \frac{1}{k}\frac{\sinh kt}{\sinh kl}\int_t^l g(s)\sinh[k(l-s)]\,ds. \end{aligned}$$

## 2.2 Systems of Equations

The Laplace transform method may also be applied to systems of linear differential equations in several unknown functions. The method of procedure in the normal case is shown in the next two examples.

**E2.2.1**

$$\frac{\mathrm{d}u}{\mathrm{d}t} = 2u + v + t,$$

$$\frac{\mathrm{d}v}{\mathrm{d}t} = u + 2v - 1,$$

with $u(0) = 0$, $v(0) = 1$.

The transforms of the equations are

$$p\bar{u} = 2\bar{u} + \bar{v} + p^{-2},$$
$$p\bar{v} - 1 = \bar{u} + 2\bar{v} - p^{-1},$$

that is

$$(p-2)\bar{u} - \bar{v} = p^{-2},$$
$$-\bar{u} + (p-2)\bar{v} = 1 - p^{-1}.$$

Solving for $\bar{u}$ and $\bar{v}$, we obtain

$$\bar{u}(p) = \frac{p^2 - 2}{p^2[(p-2)^2 - 1]} = \frac{7}{18(p-3)} + \frac{1}{2(p-1)} - \frac{8}{9p} - \frac{2}{3p^2},$$

$$\bar{v}(p) = \frac{p^3 - 3p^2 + 2p + 1}{p^2[(p-2)^2 - 1]} = \frac{7}{18(p-3)} - \frac{1}{2(p-1)} + \frac{10}{9p} + \frac{1}{3p^2}.$$

The required solution is therefore

$$u(t) = \tfrac{7}{18}\mathrm{e}^{3t} + \tfrac{1}{2}\mathrm{e}^{t} - \tfrac{8}{9} - \tfrac{2}{3}t,$$
$$v(t) = \tfrac{7}{18}\mathrm{e}^{3t} - \tfrac{1}{2}\mathrm{e}^{t} + \tfrac{10}{9} + \tfrac{1}{3}t.$$

**E2.2.2** A typical problem of small oscillations:

$$(\mathrm{D}^2 + 2)u - v = 0,$$
$$-2u + (\mathrm{D}^2 + 3)v = 0,$$

with $u(0) = 1$, $u'(0) = v(0) = v'(0) = 0$.

Here

$$(p^2 + 2)\bar{u} - \bar{v} = p,$$
$$-2\bar{u} + (p^2 + 3)\bar{v} = 0,$$

which give

$$\bar{u}(p) = \frac{p(p^2+3)}{(p^2+2)(p^2+3)-2} = \frac{2p}{3(p^2+1)} + \frac{p}{3(p^2+4)},$$

$$\bar{v}(p) = \frac{2p}{(p^2+2)(p^2+3)-2} = \frac{2p}{3(p^2+1)} - \frac{2p}{3(p^2+4)}.$$

Thus

$$u(t) = \tfrac{2}{3}\cos t + \tfrac{1}{3}\cos 2t,$$
$$v(t) = \tfrac{2}{3}\cos t - \tfrac{2}{3}\cos 2t.$$

Each of the examples E2.2.1 and E2.2.2 contains a *normal* system of equations, that is one in which all the initial conditions needed to determine the transforms of the equations can be chosen arbitrarily. Any normal system can be expressed as a system of first-order equations with an initial condition for each unknown function, by introducing additional variables to denote the intermediate derivatives. For example, E2.2.2 can be written as

$$\begin{aligned} 2u - v \quad + \mathrm{D}w \qquad &= 0, \\ -2u + 3v \qquad\qquad + \mathrm{D}x &= 0, \\ \mathrm{D}u \qquad\quad - w \qquad &= 0, \\ \mathrm{D}v \qquad\quad - x &= 0, \end{aligned}$$

with $u(0) = 1$, $v(0) = w(0) = x(0) = 0$.

A system of equations need not be normal, as the solution of the following example contains no arbitrary constant.

**E2.2.3**

$$\mathrm{D}u + (\mathrm{D}+1)v = 1, \quad (\mathrm{D}-1)u + \mathrm{D}v = 0.$$

If $u(0) = u_0$, $v(0) = v_0$, the equations for the transforms are

$$p\bar{u} + (p+1)\bar{v} = p^{-1} + u_0 + v_0,$$
$$(p-1)\bar{u} + p\bar{v} = u_0 + v_0.$$

On subtraction we have $\bar{u} + \bar{v} = p^{-1}$, so that we obtain

$$\bar{u}(p) = 1 - u_0 - v_0,$$
$$\bar{v}(p) = p^{-1} - 1 + u_0 + v_0.$$

Since $\bar{u} + \bar{v} = p^{-1}$, $u + v = 1$, and since $u \to u_0$, $v \to v_0$ as $t \to 0$ we must have $u_0 + v_0 = 1$. Consequently, $\bar{u}(p) = 0$, $\bar{v}(p) = p^{-1}$ and the required solution is $u(t) = 0$, $v(t) = 1$.

**P2.2.1** Consider the general system of equations

$$\sum_{j=1}^{n} P_{ij}(\mathrm{D})u_j(t) = g_i(t) \qquad (i = 1, 2, \ldots, m), \tag{2.2.1}$$

where the differential operators $P_{ij}(\mathrm{D})$ are polynomials in $\mathrm{D} \equiv \mathrm{d}/\mathrm{d}t$ with constant coefficients, the functions $u_j(t)$ are the unknowns, and the functions $g_i(t)$ are given. We can eliminate any of the functions $u_j(t)$ between pairs of the equations (2.2.1) as if these were algebraic equations, because the operators $P_{ij}(\mathrm{D})$ commute. Consequently, the system (2.2.1) may be underdetermined, exactly determined, or inconsistent, in the same way as a system of algebraic equations, provided that suitable initial conditions are given.

If $m = n$, so that the system (2.2.1) has as many equations as unknowns, the result of eliminating all but one of the unknown functions is of the form

$$\Delta(\mathrm{D})u_j(t) = \sum_{i=1}^{n} Q_{ij}(\mathrm{D})g_i(t), \tag{2.2.2}$$

where $\Delta(\mathrm{D}) = \det(P_{ij}(\mathrm{D}))$ and $Q_{ij}(\mathrm{D})$ is the cofactor of $P_{ij}(\mathrm{D})$ in $\Delta(\mathrm{D})$. The order of equation (2.2.2) is the degree $N$ of the polynomial $\Delta(\mathrm{D})$, and this is the number of the arbitrary constants that appear in the general solution of (2.2.1).

To show this, and to obtain the general solution in a systematic way, the elimination of the functions $u_j(t)$ should be done so as to reduce (2.2.1) to upper-triangular form, from which the functions can be determined successively by back-substitution. The elimination is carried out by the highest-common-factor process, as shown in the following example.

**P2.2.4**

$$(\mathrm{D}^3 - 1)u + (\mathrm{D}^2 + 1)v + \mathrm{D}w = 1, \tag{a}$$
$$(\mathrm{D}^2 - 1)u + (\mathrm{D} + 1)v + w = t, \tag{b}$$
$$(\mathrm{D} - 1)u + v = t^2. \tag{c}$$

The determinant of the system is $\Delta(\mathrm{D}) = -\mathrm{D}(\mathrm{D} - 1)$, so that it is of second order. Replace the equations in turn as follows.

$$(\mathrm{a}) - \mathrm{D}(\mathrm{b}):\quad (\mathrm{D} - 1)u + (-\mathrm{D} + 1)v = 0, \tag{a$'$}$$
$$(\mathrm{b}) - \mathrm{D}(\mathrm{c}):\quad (\mathrm{D} - 1)u + v + w = -t, \tag{b$'$}$$
$$(\mathrm{a}') - (\mathrm{b}'):\quad -\mathrm{D}v - w = t, \tag{a$''$}$$
$$(\mathrm{c}) - (\mathrm{b}'):\quad -w = t^2 + t. \tag{c$'$}$$

The original system is equivalent to the upper-triangular system (b′), (a″), (c′). This is solved in the reverse order to give

$$w = -t^2 - t, \qquad v = \tfrac{1}{3}t^3 + A, \qquad u = \tfrac{1}{3}t^3 + A + Be^t,$$

where $A$ and $B$ are the two arbitrary constants.

**P2.2.1** (continued) The elimination process, carried out as in E2.2.4, leaves the determinant $\Delta(\mathrm{D})$ unaltered apart from sign. As shown in E2.2.4, the final system may contain both differential and algebraic equations for the unknown functions, but when they are solved in turn the number of arbitrary constants introduced is equal to the sum of the orders of the terms in the leading diagonal, and so to the degree $N$ of $\Delta(\mathrm{D})$.

E2.2.4 also shows that the arbitrary constants do not necessarily appear in each of the functions. Thus $w$ contains neither of the constants, and $B$ appears only in $u$, so that no initial condition can be laid on $w$ and at least one must be put on $u$, in order to determine the solution. The second condition may be on either $u$ or $v$. For any system, the greatest number of initial conditions that can be placed on a function is found by putting that function in the last column of the upper-triangular set of equations, and the least number of conditions by putting the function in the first column. A full discussion of these questions is given by Ince (1927, Sections 6.4 to 6.53).

As noted earlier, a normal system can be expressed as

$$\sum_{j=1}^{N} (a_{ij}\mathrm{D} + b_{ij})u_j(t) = g_i(t) \qquad (i = 1, 2, \ldots, N),$$

$$u_j(0) = c_j \qquad (j = 1, 2, \ldots, N),$$

with $\det(a_{ij}) \neq 0$. The Laplace transform of this system is

$$\sum_{j=1}^{N} (a_{ij}p + b_{ij})\bar{u}_j(p) = \bar{g}_i(p) + \sum_{j=1}^{N} a_{ij}c_j,$$

so that

$$\Delta(p)\bar{u}_j(p) = \sum_{i=1}^{N} Q_{ij}(p)\left(\bar{g}_i(p) + \sum_{k=1}^{N} a_{ik}c_k\right).$$

The cofactors $Q_{ij}(p)$ are polynomials of degree $N-1$ at most, whereas $\Delta(p)$ is of degree $N$. The solution can therefore be written as

$$u_j(t) = \sum_{i=1}^{N} q_{ij}(t) * g_i(t) + \sum_{i,k=1}^{N} a_{ik}c_k q_{ij}(t),$$

where $q_{ij}(t)$ is the e.t.f. with Laplace transform

$$\bar{q}_{ij}(p) = \frac{Q_{ij}(p)}{\Delta(p)}.$$

A well posed physical problem should lead to a normal system of equations and initial conditions. If a system such as that of E2.2.4 has arbitrary forcing functions on its right-hand side, then when it is reduced to a normal system the derivatives of these functions may appear on the right-hand side, and also in the solution.

## 2.3 Electric Circuit Problems

The problems of transients in electric circuits provide an important and interesting field of application for Laplace transforms. The circuits to be considered are made up of three basic types of element, having the properties of resistance ($R$), inductance ($L$) and capacitance ($C$), which are represented conventionally as shown in Figure 2.1.

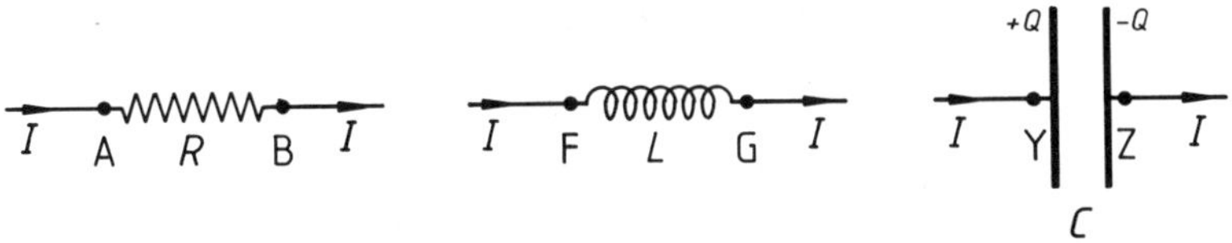


Figure 2.1

These elements have the properties that the differences of the electric potential between their two sides are given by the equations

$$V_A - V_B = RI, \qquad V_F - V_G = L\frac{dI}{dt}, \qquad V_Y - V_Z = \frac{Q}{C}.$$

Here $t$ denotes time, $I$ is the current entering and leaving the elements in the directions shown, and $\pm Q$ are the charges carried by the two plates of the condenser or capacitance element. Since the current in a circuit is carried by the flow of charge, the current and charge are related by

$$I = \frac{dQ}{dt}.$$

It will be assumed that the quantities $R$, $L$ and $C$ are constant for any element, and the standard units of measurement are as follows:

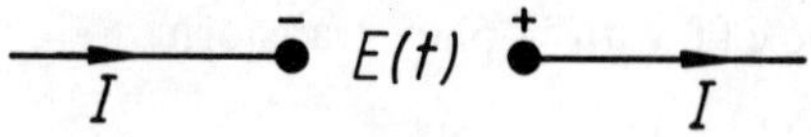


Figure 2.2

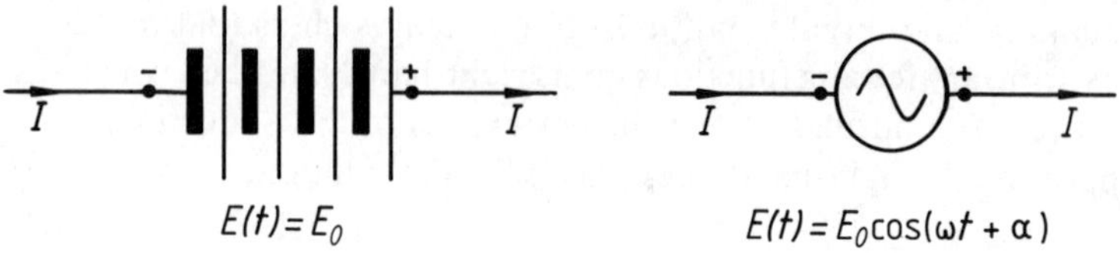


Figure 2.3

for time the second; potential difference the volt; charge the coulomb; current the ampere; resistance the ohm; inductance the henry; capacity the farad. The equations are in dimensionally correct form, so that they can be used with any consistent set of units.

When the elements are connected in a circuit, a current may be caused to flow by the application of an electromotive force (e.m.f.) which produces a potential difference. This is usually taken to be either constant, such as from a battery of electrolytic cells, or varying sinusoidally with time, such as from a generator. In general, the e.m.f. will be denoted by $E(t)$ and represented as shown in Figure 2.2. The positive direction of current flow through the circuit is from the positive terminal to the negative one. In the two particular cases mentioned, the representations shown in Figure 2.3 are used.

**E2.3.1** The simple circuit composed of a resistance, an inductance and a capacitance connected in series, to which is applied the e.m.f. $E(t)$, as shown in Figure 2.4.

Since the differences of potential across the elements can be

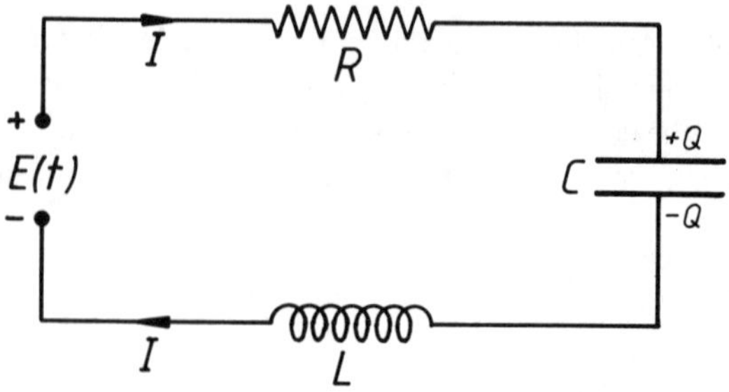


Figure 2.4

added, the equations for the circuit are

$$L\frac{dI}{dt}+RI+\frac{Q}{C}=E(t), \qquad \frac{dQ}{dt}=I.$$

The Laplace transforms of these equations are

$$L[p\bar{I}-I(0)]+R\bar{I}+\bar{Q}/C=\bar{E}(p), \qquad p\bar{Q}-Q(0)=\bar{I},$$

where $Q(0)$ and $I(0)$ are the charge and current at time $t=0$.

We are usually interested in the current, so we eliminate $\bar{Q}$ to obtain

$$\left(Lp+R+\frac{1}{Cp}\right)\bar{I}(p)=\bar{E}(p)+LI(0)-\frac{Q(0)}{Cp}.$$

The coefficient of $\bar{I}(p)$, namely

$$Z(p)=Lp+R+\frac{1}{Cp},$$

is called the *generalized impedance* of the circuit. If the circuit is initially dead, that is $I(0)=Q(0)=0$, we have

$$\bar{I}(p)=\bar{E}(p)/Z(p),$$

a relation which generalises the simple Ohm's law for the resistance $R$. The equation for $\bar{I}(p)$ also shows that an initial charge is equivalent to an e.m.f. $-Q(0)C^{-1}\mathrm{H}(t)$ and an initial current to an e.m.f. $LI(0)\delta(t)$.

In particular, if $E(t)=E_0\cos(\omega t+\alpha)$ we have

$$\bar{E}(p)=E_0\frac{p\cos\alpha-\omega\sin\alpha}{p^2+\omega^2},$$

so that if $I(0)=Q(0)=0$

$$\bar{I}(p)=E_0\frac{p\cos\alpha-\omega\sin\alpha}{(p^2+\omega^2)Z(p)}.$$

Now $Z(p)=L(p-\lambda)(p-\mu)/p$, where $\lambda$ and $\mu$ are the roots of the quadratic equation

$$LCp^2+RCp+1=0.$$

Since $L$, $C$ and $R$ are all positive, $\lambda$ and $\mu$ are either real and negative or conjugate complex numbers with a negative real part. Provided that $\lambda \neq \mu$, the partial-fraction expression for $\bar{I}(p)$ is

$$\bar{I}(p) = \frac{E_0 e^{i\alpha}}{2Z(i\omega)(p - i\omega)} + \frac{E_0 e^{-i\alpha}}{2Z(-i\omega)(p + i\omega)} + \frac{E_0}{L(\lambda - \mu)}\left(\frac{\lambda^2 \cos\alpha - \lambda\omega \sin\alpha}{(\lambda^2 + \omega^2)(p - \lambda)} - \frac{\mu^2 \cos\alpha - \mu\omega \sin\alpha}{(\mu^2 + \omega^2)(p - \mu)}\right).$$

The current is therefore

$$I(t) = A_s \cos(\omega t + \alpha + \theta) + A_1 e^{\lambda t} + A_2 e^{\mu t},$$

where $A_s$ is the amplitude of the ultimate steady oscillation, and the last two terms are the *transients*, since they decay as $t \to \infty$. Here

$$A_s e^{i\theta} = \frac{E_0}{Z(i\omega)} = \frac{E_0}{R + [L\omega - (1/C\omega)]i},$$

and $A_1$ and $A_2$ are the coefficients of $(p - \lambda)^{-1}, (p - \mu)^{-1}$ in $\bar{I}(p)$. The transients are solutions of the circuit equations without any applied e.m.f., and are oscillatory if $\lambda$ and $\mu$ are complex, that is if $4L > R^2C$. The rates of decay depend only on $R$, $L$ and $C$, but the amplitudes of the transients depend on the applied e.m.f. $E(t)$ and on the initial conditions.

From the basic elements $R, L$ and $C$, more complicated circuits can be built up by linking these elements in the series or parallel, and these circuits can be analysed by means of *Kirchhoff's laws*:

1. There is no accumulation of charge at a junction, so that the sum of the currents flowing into the junction is equal to the sum of the currents flowing out.
2. The total fall of potential round any closed loop in the circuit is zero.

These laws imply that the rules for combining generalized impedances in series or parallel are the same as for resistances (Figure 2.5).

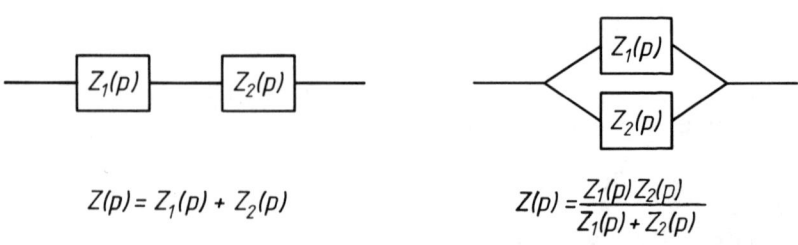

$Z(p) = Z_1(p) + Z_2(p)$ $\qquad$ $Z(p) = \dfrac{Z_1(p)Z_2(p)}{Z_1(p) + Z_2(p)}$

Figure 2.5

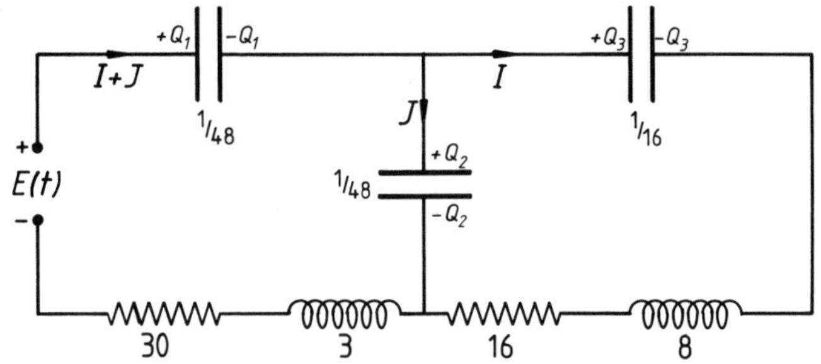


Figure 2.6

**E2.3.2** Let $I$ and $J$ be the currents in the two branches shown in Figure 2.6. By Kirchhoff's first law, the total current to and from the e.m.f. $E(t)$ is $I + J$. The second law then gives the circuit equations for the two loops as

$$3\frac{\mathrm{d}}{\mathrm{d}t}(I+J) + 30(I+J) + 48_1 + 48_2 = E(t),$$

$$8\frac{\mathrm{d}I}{\mathrm{d}t} + 16I + 16Q_3 - 48Q_2 = 0.$$

The charges and currents are related by

$$\frac{\mathrm{d}Q_1}{\mathrm{d}t} = I + J, \qquad \frac{\mathrm{d}Q_2}{\mathrm{d}t} = J, \qquad \frac{\mathrm{d}Q_3}{\mathrm{d}t} = I.$$

If the circuit is initially dead, the Laplace transforms of these equations are

$$(3p+30)(\bar{I}+\bar{J}) + 48(\bar{Q}_1 + \bar{Q}_2) = \bar{E}(p),$$

$$(8p+16)\bar{I} + 16\bar{Q}_3 - 48\bar{Q}_2 = 0,$$

$$p\bar{Q}_1 = \bar{I} + \bar{J}, \qquad p\bar{Q}_2 = \bar{J}, \qquad p\bar{Q}_3 = \bar{I}.$$

The transforms of the currents are therefore given by

$$(p^2 + 10p + 16)\bar{I} + (p^2 + 10p + 32)\bar{J} = \tfrac{1}{3}p\bar{E}(p),$$

$$(p^2 + 2p + 2)\bar{I} - 6\bar{J} = 0,$$

so that

$$\bar{I}(p) = \frac{2p\bar{E}(p)}{(p^2+4p+8)(p^2+8p+20)},$$

$$\bar{J}(p) = \frac{p(p^2+2p+2)\bar{E}(p)}{3(p^2+4p+8)(p^2+8p+20)}.$$

The factors in the denominators of $\bar{I}$ and $\bar{J}$ correspond to the existence of transients which are decaying free oscillations of the forms

$$A_1 e^{-2t} \cos(2t + \delta_1) \qquad \text{and} \qquad A_2 e^{-4t} \cos(2t + \delta_2).$$

If the applied e.m.f. $E(t)$ is sinusoidal, there will be forced oscillations in $I(t)$ and $J(t)$ with the same period as $E(t)$. Let

$$E(t) = E_0 \cos(\omega t + \alpha),$$

so that

$$\bar{E}(p) = E_0 \frac{p \cos \alpha - \omega \sin \alpha}{p^2 + \omega^2} = \tfrac{1}{2} E_0 \left( \frac{e^{i\alpha}}{p - i\omega} + \frac{e^{-i\alpha}}{p + i\omega} \right).$$

The transients and forced oscillations are found by expressing $\bar{I}(p)$ and $\bar{J}(p)$ in partial fractions. In $\bar{I}(p)$, the coefficient of $(p - i\omega)^{-1}$ is

$$\tfrac{1}{2} E_0 e^{i\alpha} \frac{2i\omega}{(-\omega^2 + 4i\omega + 8)(-\omega^2 + 8i\omega + 20)}$$

and the coefficient of $(p + i\omega)^{-1}$ is the complex conjugate of this. Hence the forced steady oscillation in $I(t)$ is

$$I_s(t) = A_s \cos(\omega t + \alpha + \theta),$$

where

$$A_s e^{i\theta} = \frac{2i\omega E_0}{(8 - \omega^2 + 4i\omega)(20 - \omega^2 + 8i\omega)}.$$

Similarly, the coefficient of $(p + 2 - 2i)^{-1}$ in $\bar{I}(p)$ is

$$\frac{2(-2 + 2i)[(-2 + 2i) \cos \alpha - \omega \sin \alpha] E_0}{[(-2 + 2i)^2 + \omega^2] 4i(4 + 8i)},$$

and that of $(p + 4 - 2i)^{-1}$ is

$$\frac{2(-4 + 2i)[(-4 + 2i) \cos \alpha - \omega \sin \alpha] E_0}{[(-4 + 2i)^2 + \omega^2](4 - 8i) 4i},$$

so that the transient part of $I(t)$ is

$$I_u(t) = A_1 e^{-2t} \cos(2t + \phi_1) + A_2 e^{-4t} \cos(2t + \phi_2),$$

where

$$A_1 \exp(i\phi_1) = \frac{(3 - i)[(-2 + 2i) \cos \alpha - \omega \sin \alpha] E_0}{10(\omega^2 - 8i)},$$

$$A_2 \exp(i\phi_2) = \frac{(-3 + 4i)[(-4 + 2i) \cos \alpha - \omega \sin \alpha] E_0}{10(\omega^2 + 12 - 16i)}.$$

In the same way, it is found that $J(t) = J_s(t) + J_u(t)$, where

$$J_s(t) = B_s \cos(\omega t + \alpha + \lambda),$$
$$J_u(t) = B_1 e^{-2t} \cos(2t + \mu_1) + B_2 e^{-4t} \cos(2t + \mu_2),$$

and

$$B_s \exp(i\lambda) = \frac{i\omega(2 - \omega^2 + 2i\omega)E_0}{3(8 - \omega^2 + 4i\omega)(20 - \omega^2 + 8i\omega)},$$

$$B_1 \exp(i\mu_1) = \frac{4\cos\alpha + (1 + i)\omega \sin\alpha}{24(\omega^2 - 8i)} E_0,$$

$$B_2 \exp(i\mu_2) = \frac{-6i\cos\alpha - (1 + 2i)\omega\sin\alpha}{8(\omega^2 + 12 - 16i)} E_0.$$

Electric circuits may also be coupled by *mutual induction* between two coils (see Figure 2.7). In this case, the potential differences are given by

$$V_A - V_B = L_1 \frac{dI_1}{dt} + M\frac{dI_2}{dt},$$

$$V_C - V_D = M\frac{dI_1}{dt} + L_2\frac{dI_2}{dt}.$$

Here $M$ is the mutual inductance between the coils of self-inductances $L_1$ and $L_2$. $M$ may have either sign, depending on the way the coils are wound, but it is restricted by the condition

$$M^2 \leqslant L_1 L_2,$$

with equality only in the ideal limiting case of perfect coupling. Mutual inductance is the basis of the transformer circuit, treated in the next example.

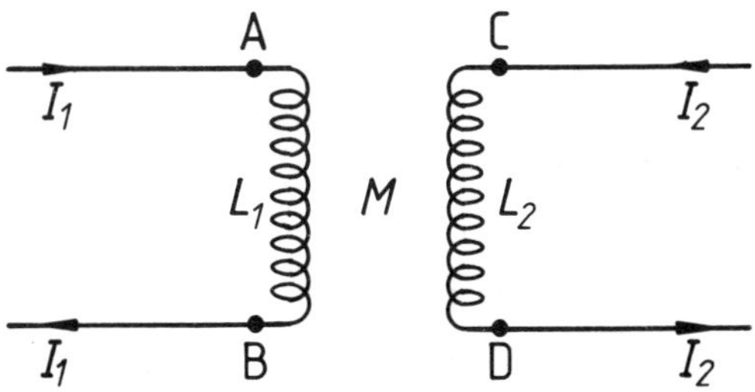


Figure 2.7

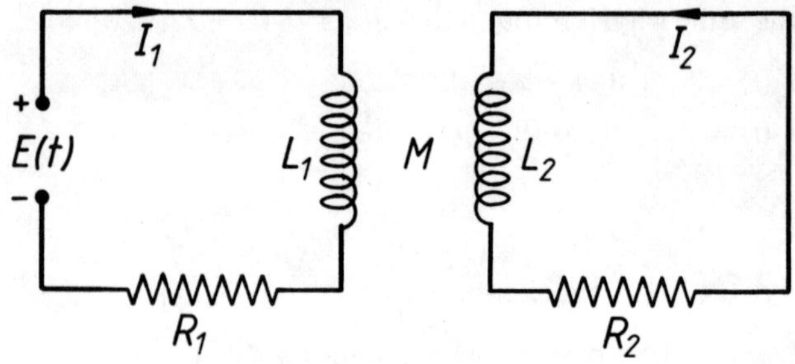


Figure 2.8

**E2.3.3** An electromotive force $E(t)$ is applied to the primary circuit and mutual induction drives the current $I_2$ in the secondary circuit of Figure 2.8. For simplicity, we shall suppose that there is no capacitance in either circuit. The circuit equations are then

$$L_1\frac{\mathrm{d}I_1}{\mathrm{d}t} + M\frac{\mathrm{d}I_2}{\mathrm{d}t} + R_1 I_1 = E(t),$$

$$M\frac{\mathrm{d}I_1}{\mathrm{d}t} + L_2\frac{\mathrm{d}I_2}{\mathrm{d}t} + R_2 I_2 = 0.$$

If there are no currents initially, the transforms of these equations are

$$(L_1 p + R_1)\bar{I}_1 + Mp\bar{I}_2 = \bar{E}(p),$$
$$Mp\bar{I}_1 + (L_2 p + R_2)\bar{I}_2 = 0,$$

from which

$$\bar{I}_2(p) = -\frac{Mp\bar{E}(p)}{(L_1 p + R_1)(L_2 p + R_2) - M^2 p^2}.$$

The important case is when $E(t) = E_0 \cos(\omega t + \alpha)$, so that

$$\bar{I}_2(p) = -\frac{ME_0 p(p\cos\alpha - \omega\sin\alpha)}{(p^2 + \omega^2)[(L_1 L_2 - M^2)p^2 + (L_1 R_2 + L_2 R_1)p + R_1 R_2]}.$$

If $L_1 L_2 > M^2$, the second factor in the denominator has two real negative zeros $p = \lambda$ and $p = \mu$, so that we can write

$$\bar{I}_2(p) = -\frac{ME_0 p(p\cos\alpha - \omega\sin\alpha)}{(L_1 L_2 - M^2)(p^2 + \omega^2)(p - \lambda)(p - \mu)}.$$

After expressing $\bar{I}_2(p)$ in partial fractions, we find that

$$I_2(t) = A_s \cos(\omega t + \alpha + \theta) + A_1 \mathrm{e}^{\lambda t} + A_2 \mathrm{e}^{\mu t},$$

where

$$A_s e^{i\theta} = -\frac{i\omega M E_0}{(i\omega L_1 + R_1)(i\omega L_2 + R_2) + M^2\omega^2},$$

$$A_1 = \frac{\lambda(\lambda \cos\alpha - \omega \sin\alpha) M E_0}{(\lambda^2 + \omega^2)[(L_1 R_2 - L_2 R_1)^2 + 4M^2 R_1 R_2]^{1/2}},$$

$$A_2 = -\frac{\mu(\mu \cos\alpha - \omega \sin\alpha) M E_0}{(\mu^2 + \omega^2)[(L_1 R_2 - L_2 R_1)^2 + 4M^2 R_1 R_2]^{1/2}},$$

and we have supposed that $\lambda < \mu$. The first term is the forced steady oscillation in the secondary; the last two are transients since $\lambda$ and $\mu$ are negative.

When $M^2 = L_1 L_2$ the circuit equations are not normal, as they form a first-order system, and we cannot expect to be able to impose the two conditions $I_1(0) = I_2(0) = 0$. It is therefore of interest to see what happens to the solution already found as $M^2 \to L_1 L_2$. In this limit one of the decay rates remains finite, as

$$\mu \to -\frac{R_1 R_2}{L_1 R_2 + L_2 R_1}, \qquad \text{but} \qquad \lambda \sim -\frac{L_1 R_2 + L_2 R_1}{L_1 L_2 - M^2} \to -\infty.$$

Hence, we find that as $M \to \sqrt{(L_1 L_2)}$, $A_s \to A \sin\beta$, $\theta \to -(\frac{1}{2}\pi + \beta)$, $A_1 \to A \cos\alpha$ and $A_2 \to -A\cos(\alpha - \beta)\cos\beta$, where

$$A = \frac{M E_0}{L_1 R_2 + L_2 R_1} \qquad \text{and} \qquad \tan\beta = \omega\left(\frac{L_1}{R_1} + \frac{L_2}{R_2}\right).$$

Because $\lambda$ is large and negative when $L_1 L_2 - M^2$ is small, the term $A_1 e^{\lambda t}$ represents a very rapidly decaying transient. Thus, unless $\cos\alpha = 0$, the current $I_2(0)$ changes very rapidly from 0 to $-A\cos\alpha$, which may be much greater in magnitude than the final steady amplitude $A \sin\beta$ if $\omega$ is small. It can similarly be shown that there is also a rapid change in $I_1(t)$.

These results can be obtained directly by setting $M^2 = L_1 L_2$ in the transforms $\bar{I}_2(p)$ and $\bar{I}_1(p)$. This gives, from E1.1.7,

$$I_2(0) = -\frac{M E_0 \cos\alpha}{L_1 R_2 + L_2 R_1} \qquad \text{and} \qquad I_1(0) = \frac{L_2 E_0 \cos\alpha}{L_1 R_2 + L_2 R_1},$$

even though these transforms were derived on the assumption that $I_1(0) = I_2(0) = 0$. The inconsistency arises because we are no longer dealing with a normal system of equations.

## 2.4 Equations with Polynomial Coefficients

The Laplace transformation is less helpful for the solution of differential equations when the coefficients are not constant. One reason is that the solution does not necessarily have a Laplace transform. Assuming that it does exist, we can transform an equation with polynomial coefficients by means of P1.1.9, but the result is now a differential equation for the transform. It is worth noting that

$$\mathscr{L}\left\{t\frac{\mathrm{d}u}{\mathrm{d}t}\right\} = \mathscr{L}\left\{\frac{\mathrm{d}}{\mathrm{d}t}(tu) - u\right\} = -p\frac{\mathrm{d}\bar{u}}{\mathrm{d}p} - \bar{u}$$

does not involve the assumption that $u(0)$ is finite.

**E2.4.1**

$$t\frac{\mathrm{d}^2u}{\mathrm{d}t^2} + u = 0, \qquad u(0) = 0.$$

The Laplace transform of the equation is

$$-p^2\frac{\mathrm{d}\bar{u}}{\mathrm{d}p} - 2p\bar{u} + \bar{u} = 0,$$

so that

$$\bar{u}(p) = Cp^{-2}\exp(-p^{-1}),$$

where $C$ is a constant. From E1.3.1 we see that

$$u(t) = Ct^{1/2}\mathrm{J}_1(2t^{1/2}).$$

Differential equations are sometimes useful for the evaluation of the transforms of known functions.

**E2.4.2** $u(t) = t^{-1}\sin t$ satisfies

$$\frac{\mathrm{d}^2}{\mathrm{d}t^2}(tu) + tu = 0, \qquad u(0) = 1.$$

The transform of this equation is

$$-p^2\frac{\mathrm{d}\bar{u}}{\mathrm{d}p} - 1 - \frac{\mathrm{d}\bar{u}}{\mathrm{d}p} = 0,$$

so that

$$\frac{\mathrm{d}\bar{u}}{\mathrm{d}p} = -\frac{1}{p^2+1}.$$

Integration gives $\bar{u}(p) = C - \tan^{-1} p$, where $C$ is a constant. From P1.2.8 $\bar{u}(p) \to 0$ as $\mathscr{R} p \to \infty$, so that $C = \frac{1}{2}\pi$ and $\bar{u}(p) = \cot^{-1} p$, as found in E1.3.4.

**E2.4.3** The Bessel functions $J_\nu(t)$ and $Y_\nu(t)$ satisfy

$$t^2 \frac{d^2 u}{dt^2} + t \frac{du}{dt} + (t^2 - \nu^2) u = 0.$$

The transform of this equation is

$$(p^2 + 1) \frac{d^2 \bar{u}}{dp^2} + 3p \frac{d\bar{u}}{dp} + (1 - \nu^2) \bar{u} = 0,$$

which can be expressed as

$$\frac{d}{dp}\left( (p^2 + 1) \frac{d\bar{u}}{dp} + p\bar{u} \right) = \nu^2 \bar{u},$$

that is

$$\frac{d}{dp}\left( (p^2 + 1)^{1/2} \frac{d}{dp}[(p^2 + 1)^{1/2} \bar{u}] \right) = \nu^2 \bar{u}.$$

Hence if $p = \sinh z$ the equation becomes

$$\frac{d^2}{dz^2}(\bar{u} \cosh z) = \nu^2 \bar{u} \cosh z,$$

so that

$$\begin{aligned} \bar{u}(p) &= (A e^{\nu z} + B e^{-\nu z}) \operatorname{sech} z \\ &= \frac{A[(p^2 + 1)^{1/2} + p]^\nu + B[(p^2 + 1)^{1/2} + p]^{-\nu}}{(p^2 + 1)^{1/2}}. \end{aligned}$$

Now

$$J_\nu(t) = \frac{(\frac{1}{2} t)^\nu}{\Gamma(\nu + 1)} + O(t^{\nu + 2})$$

as $t \to 0$, so that from P1.2.13

$$\mathscr{L}\{J_\nu(t)\} \sim 2^{-\nu} p^{-\nu - 1} \qquad \text{as } \mathscr{R} p \to +\infty,$$

provided $\mathscr{R} \nu > -1$. Hence we must have

$$\mathscr{L}\{J_\nu(t)\} = (p^2 + 1)^{-1/2} [(p^2 + 1)^{1/2} + p]^{-\nu}$$

for $\mathscr{R} \nu > -1$, as stated in E1.3.2.

The second solution $Y_\nu(t)$ also has a Laplace transform if $-1 < \mathcal{R}\nu < 1$, and from D1.3.1

$$\mathcal{L}\{Y_\nu(t)\} = (p^2+1)^{-1/2}\{\cot \nu\pi[(p^2+1)^{1/2}+p]^{-\nu} - \operatorname{cosec} \nu\pi[(p^2+1)^{1/2}+p]^{\nu}\}.$$

This fails when $\nu = 0$, in which case

$$\bar{u}(p) = (Az+B)\operatorname{sech} z = (p^2+1)^{-1/2}(A\sinh^{-1} p + B).$$

Since

$$Y_0(t) = \lim_{\nu\to 0} Y_\nu(t)$$

we get

$$\mathcal{L}\{Y_0(t)\} = -\frac{2z}{\pi\cosh z} = -\frac{2}{\pi}(p^2+1)^{-1/2}\sinh^{-1} p.$$

**E2.4.4** Airy's equation is

$$\frac{d^2u}{dt^2} = tu.$$

If the solution $u(t)$ has a Laplace transform, $\bar{u}(p)$ satisfies

$$p^2\bar{u} - pu_0 - u_1 = -\frac{d\bar{u}}{dp},$$

where $u_0 = u(0)$, $u_1 = u'(0)$. The general solution of this equation is

$$\bar{u}(p) = \exp(-\tfrac{1}{3}p^3)\left(\int_0^p \exp(\tfrac{1}{3}z^3)(zu_0+u_1)\,dz + \bar{u}(0)\right).$$

From P1.2.8 $\bar{u}(p)\to 0$ as $|p|\to\infty$ with $|\arg p| \leqslant \tfrac{1}{2}\pi - \delta$. This condition is satisfied automatically for $|\arg p| < \tfrac{1}{6}\pi$, but we have to impose it in the remainder of the sector. If we put

$$p = \exp(\pm\tfrac{1}{3}\pi i)q, \qquad z = \exp(\pm\tfrac{1}{3}\pi i)\zeta$$

we have

$$\bar{u}(\exp(\pm\tfrac{1}{3}\pi i)q) = \exp(\tfrac{1}{3}q^3)\left(\int_0^q \exp(-\tfrac{1}{3}\zeta^3) \times [\exp(\pm\tfrac{1}{3}\pi i)\zeta u_0 + u_1]\exp(\pm\tfrac{1}{3}\pi i)\,d\zeta + \bar{u}(0)\right).$$

Hence we obtain the two conditions

$$\bar{u}(0) = -\int_0^\infty \exp(-\tfrac{1}{3}\zeta^3)[\exp(\pm\tfrac{2}{3}\pi i)\zeta u_0 + \exp(\pm\tfrac{1}{3}\pi i)u_1]\,d\zeta,$$

so that

$$\bar{u}(0) = -\exp(\pm\tfrac{2}{3}\pi i)3^{-1/3}\Gamma(\tfrac{2}{3})u_0 - \exp(\pm\tfrac{1}{3}\pi i)3^{-2/3}\Gamma(\tfrac{1}{3})u_1,$$

for both upper and lower signs. These conditions give

$$\bar{u}(0) = 3^{-1/3}\Gamma(\tfrac{2}{3})u_0 = -3^{-2/3}\Gamma(\tfrac{1}{3})u_1.$$

The standard solution of this type is the *Airy function* $\mathrm{Ai}(t)$, defined by taking

$$\mathrm{Ai}(0) = 3^{-2/3}/\Gamma(\tfrac{2}{3}), \qquad \mathrm{Ai}'(0) = -3^{-1/3}/\Gamma(\tfrac{1}{3}).$$

Thus for this solution

$$\bar{u}(0) = \int_0^\infty \mathrm{Ai}(t)\,dt = \tfrac{1}{3}.$$

## 2.5 Additional Examples

**E2.5.1** Solve the following differential equations with the given initial conditions:

(i) $u'' - 2u' - 3u = 9t^2 + 5e^{-2t}$, $\quad u(0) = 1, u'(0) = 0$;
(ii) $u'' + 4u' + 3u = 8\cos t - 6\sin t$, $\quad u(0) = 0, u'(0) = 1$;
(iii) $u'' - 2u' + 5u = 5 - 16e^{-t}$, $\quad u(0) = 0, u'(0) = 3$;
(iv) $u'' - 3u' + 2u = e^t$, $\quad u(0) = u'(0) = 0$;
(v) $u'' + 4u' + 4u = t + e^{-2t}$, $\quad u(0) = 2, u'(0) = -2$;
(vi) $u'' - 9u = 7\sinh 3t$, $\quad u(0) = -4, u'(0) = \tfrac{2}{3}$;
(vii) $u''' + 2u'' - u' - 2u = e^{-t}$, $\quad u(0) = u'(0) = 0, u''(0) = 1$;
(viii) $u''' - u'' + u' - u = e^t$, $\quad u(0) = 0, u'(0) = 1, u''(0) = -1$;
(ix) $u^{\mathrm{iv}} - 8u'' - 9u = 10\sin t$,
$u(0) = 0, u'(0) = 1, u''(0) = 0, u'''(0) = 8$.

**E2.5.2** Solve the following systems of differential equations with the given initial conditions:
(i) $u' + 2v = 0$, $\quad v' + 3u + v = e^t$, $\quad u(0) = v(0) = 0$;
(ii) $2u' + v = \cos t$, $\quad v' - 2u = 3\sin t$, $\quad u(0) = 0, v(0) = 1$;
(iii) $u' - u + v = 0$, $\quad v' - u - v = 0$, $\quad u(0) = v(0) = 1$;
(iv) $u'' + v'' + v = 0$, $\quad u'' + 6u + 2v = \cos 2t$,
$u(0) = 1, u'(0) = 0, v(0) = 0, v'(0) = -2$;

(v) $u'' + u - 4v' = 3, \qquad - u' + v'' + v = 0,$
$u(0) = v(0) = 1, u'(0) = v'(0) = 0;$
(vi) $u'' + 2u' + u - v = 0, \qquad u + v'' + 2v' + 3v = 0,$
$u(0) = 1, u'(0) = - 1, v(0) = v'(0) = 0.$

**E2.5.3** Solve the following differential equations with the given initial conditions:

(i) $u'' + u = \begin{cases} \sin t & (0 < t < \pi), \\ 0 & (t > \pi), \end{cases} \qquad u(0) = u'(0) = 0;$

(ii) $u'' + 3u' + 2u = \begin{cases} t & (0 < t < a), \\ ae^{-(t-a)} & (t > a), \end{cases} \qquad u(0) = u'(0) = 0.$

**E2.5.4** Solve the following differential equations subject to the given conditions:
(i) $u'' - u' - 2u = e^{-t}, \qquad u(0) = 1$, $u(t)$ bounded as $t \to \infty$;
(ii) $u''' - u = - 1, \qquad u(0) = u'(0) = 0, u(t) \to 1$ as $t \to \infty$;
(iii) $u^{iv} + 4u = 0, \qquad u(0) = 1, u'(0) = - 1$, $u(t)$ bounded as $t \to \infty$;
(iv) $u'' + u = 1, \qquad u(0) = u(T) = 0;$
(v) $u'' + u = \sin t, \qquad u(0) = u(\pi), u'(0) = u'(\pi).$

**E2.5.5** The function $u(t)$ satisfies the equation $u^{iv} - u = 0$ with the conditions $u(0) = u'(0) = 0, u''(0) = 1, u(T) = 0$. Find $u(t)$ and show that

$$u'(T) = \frac{\cosh T \cos T - 1}{\sinh T - \sin T}.$$

**E2.5.6** At time $t = 0$, when the charge $Q$ is 1 and there is no current in the circuit shown in Figure 2.9, the switch is closed and the condenser is allowed to discharge. Find its charge $Q$ at time $t$.

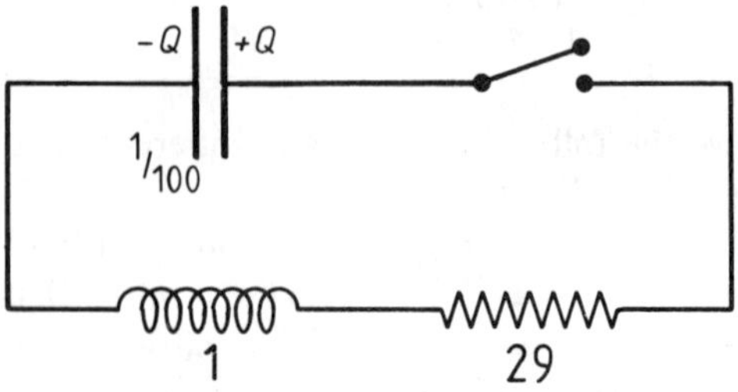


Figure 2.9

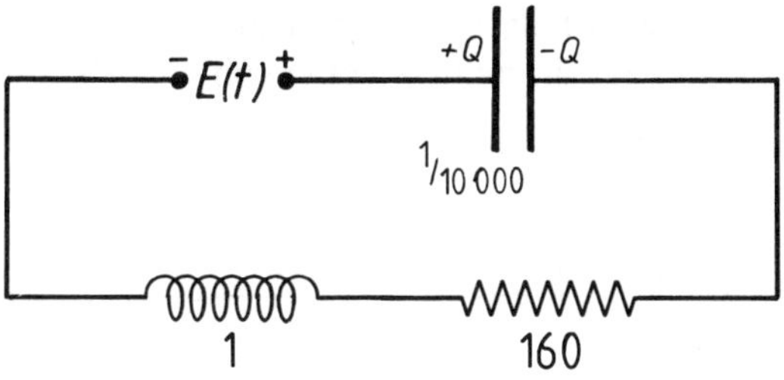


Figure 2.10

**E2.5.7** Find the charge on the condenser in the circuit shown in Figure 2.10 at time $t$, given that there is no charge or current at $t = 0$, in the two cases:
(i) $E(t) = E_0$, (ii) $E(t) = E_0 \cos(100t + \alpha)$.

**E2.5.8** Find the current through the inductance in Figure 2.11 at time $t$, given that the circuit is initially dead.

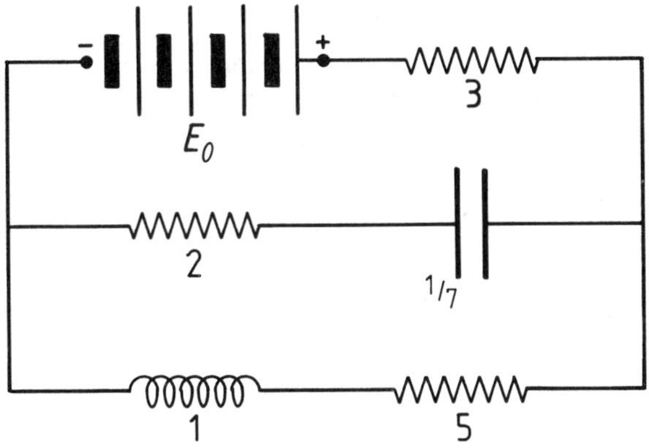


Figure 2.11

**E2.5.9** Find the charge on the condenser in Figure 2.12 at time $t$, given that the circuit is initially dead.

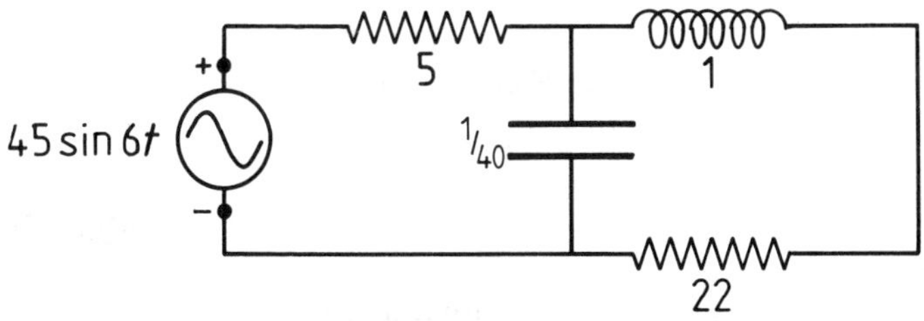


Figure 2.12

**E2.5.10** Find the current in the secondary circuit in Figure 2.13, given that there is no current in either circuit at $t = 0$.

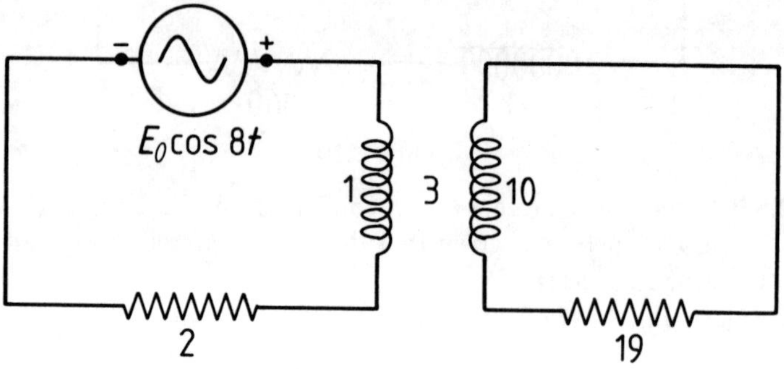


Figure 2.13

**E2.5.11** At $t = 0$, when there is no current flowing in either circuit in Figure 2.14 and the charge $Q$ is 10, the switch is closed and the condenser discharges. Find its charge $Q$ at time $t$.

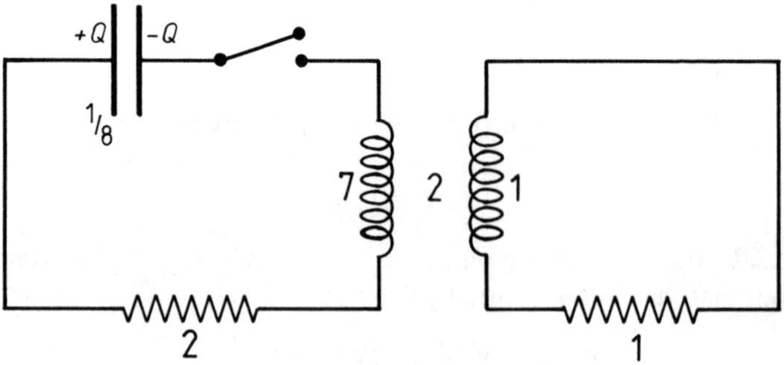


Figure 2.14

**E2.5.12** Solve the following differential equations with the given initial conditions:

(i) $(t+1)u'' + tu' - u = 0, \qquad u(0) = u'(0) = 1$;

(ii) $tu''' + tu' + 2u = 0, \qquad u(0) = 0,\ u'(0) = 1,\ u''(0) = 0.$

**E2.5.13** Use the equation $u' + 2\alpha tu = 0$ to show that (compare E1.3.6)

$$\mathscr{L}\{\exp(-\alpha t^2\} = \frac{1}{2\alpha}\int_p^\infty \exp\left(\frac{p^2 - z^2}{4\alpha}\right)\mathrm{d}z,$$

where $\mathscr{R}\alpha \geqslant 0$. Hence show that

$$\mathscr{L}\{\cos(t^2)\} = \tfrac{1}{2}\int_p^\infty \sin\left[\tfrac{1}{4}(z^2 - p^2)\right]\mathrm{d}z,$$

$$\mathscr{L}\{\sin(t^2)\} = \tfrac{1}{2}\int_p^\infty \cos\left[\tfrac{1}{4}(z^2 - p^2)\right]\mathrm{d}z.$$

**E2.5.14** Show that $u_\nu(t) = t^\nu \mathrm{J}_\nu(t)$ satisfies the equation

$$tu_\nu'' + (1 - 2\nu)u_\nu' + tu_\nu = 0,$$

and hence prove that

$$\bar{u}_\nu(p) = \frac{\Gamma(2\nu + 1)}{2^\nu\Gamma(\nu + 1)}(p^2 + 1)^{-\nu - 1/2} \qquad \text{if } \mathscr{R}\nu > 0.$$

[This result can be extended to $\mathscr{R}\nu > -\frac{1}{2}$ by using the relation $u_\nu' = tu_{\nu-1}$.]

**E2.5.15** Show that the Laplace transform of

$$u(t) = \frac{t^{a-1}}{\Gamma(a)}(1 + t)^{c-a-1},$$

where $\mathscr{R}a > 0$, satisfies the confluent hypergeometric equation

$$p\frac{\mathrm{d}^2\bar{u}}{\mathrm{d}p^2} + (c - p)\frac{\mathrm{d}\bar{u}}{\mathrm{d}p} - a\bar{u} = 0$$

and the condition $\bar{u}(p) \sim p^{-a}$ as $|p| \to \infty$ with $|\arg p| \leqslant \frac{1}{2}\pi - \delta$. [This proves that $\bar{u}(p) = U(a; c; p)$.]

**E2.5.16** If $\mathbf{A}$ is a square $n \times n$ matrix with constant elements, the exponential matrix $\mathbf{X}(t) = \exp(\mathbf{A}t)$ satisfies the equations

$$\frac{\mathrm{d}\mathbf{X}}{\mathrm{d}t} = \mathbf{AX}, \qquad \mathbf{X}(0) = \mathbf{I}.$$

Show that

$$\bar{\mathbf{X}}(p) = (p\mathbf{I} - \mathbf{A})^{-1}.$$

Use the Cayley–Hamilton theorem $f(\mathbf{A}) = \mathbf{O}$, where

$$f(\lambda) = \det(\mathbf{A} - \lambda\mathbf{I}) = \sum_{r=0}^{n} a_r \lambda^{n-r} = \prod_{i=1}^{n} (\lambda_i - \lambda),$$

to show that

$$\mathbf{X}(p) = \frac{1}{f(p)} \sum_{r=0}^{n-1} a_r \sum_{s=0}^{n-r-1} p^{n-r-s-1} \mathbf{A}^s,$$

and that $\mathbf{X}(t)$ is therefore a polynomial of degree $(n-1)$ in $\mathbf{A}$ whose coefficients are e.t.f.s. Prove that if the eigenvalues $\lambda_i$ are all distinct

$$\mathbf{X}(t) = \sum_{i=1}^{n} \exp(\lambda_i t) \prod_{j \neq i} \frac{\mathbf{A} - \lambda_j \mathbf{I}}{\lambda_i - \lambda_j}.$$

Show that the solution of the system of equations

$$\frac{d\bar{\mathbf{x}}}{dt} = \mathbf{A}\mathbf{x} + \mathbf{g}(t), \qquad \mathbf{x}(0) = \mathbf{c},$$

is

$$\mathbf{x}(t) = \mathbf{X}(t) * \mathbf{g}(t) + \mathbf{X}(t)\mathbf{c}.$$

[If $\lambda_i = \lambda$ for $1 \leqslant i \leqslant k, \lambda_j \neq \lambda$ for $j \geqslant k+1$, the contribution to $\mathbf{X}(t)$ from the zero $p = \lambda$ of $f(p)$ is

$$e^{\lambda t} \sum_{m+s \leqslant k-1} (-1)^s h_s \frac{t^m}{m!} (\mathbf{A} - \lambda\mathbf{I})^{m+s} \prod_{j=k+1}^{n} \frac{\mathbf{A} - \lambda_j \mathbf{I}}{\lambda - \lambda_j},$$

where

$$\prod_{j=k+1}^{n} \left(1 - \frac{\delta}{\lambda - \lambda_j}\right)^{-1} = \sum_{s=0}^{\infty} h_s \delta^s,$$

so that

$$h_0 = 1,$$

$$h_1 = \sum_{j=k+1}^{n} (\lambda - \lambda_j)^{-1},$$

and

$$h_s = \sum_{k+1 \leqslant j_1 \leqslant j_2 \leqslant \ldots \leqslant j_s \leqslant n} \prod_{l=1}^{s} (\lambda - \lambda_{j_l})^{-1}.]$$

**E2.5.17** Show that the equations for the currents and charges in

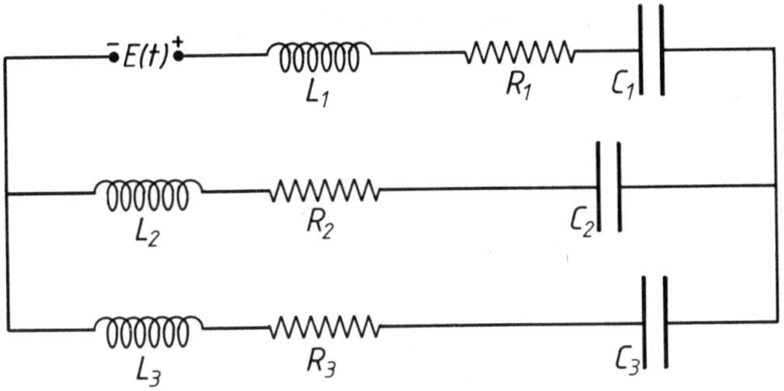


Figure 2.15

the circuit shown in Figure 2.15 form a normal system unless at least two of the inductances $L_1$, $L_2$ and $L_3$ are zero.

**E2.5.18** In the transformer circuit of E2.3.3, with $E(t) = E_0 \cos(\omega t + \alpha)$, the secondary is open when $t < 0$ so that $I_2 = 0$. Show that if the transients in the primary have disappeared

$$I_1 = \frac{E_0}{\sqrt{(R_1^2 + \omega^2 L_1^2)}} \cos(\omega t + \alpha - \theta),$$

where $\tan\theta = \omega L_1/R_1$. The secondary circuit is closed at $t = 0$, so that the initial conditions are

$$I_1(0) = \frac{E_0 \cos(\alpha - \theta)}{\sqrt{(R_1^2 + \omega^2 L_1^2)}}, \qquad I_2(0) = 0.$$

Show that when $L_1 L_2 - M^2$ is small the current in the secondary changes rapidly from 0 to the value

$$\frac{E_0 M \sin\theta \sin(\alpha - \theta)}{L_1 R_2 + L_2 R_1}.$$

**E2.5.19** Show that $u_\nu(t) = t^{-\nu} \mathrm{J}_\nu(t)$ has a Laplace transform for all values of $\nu$. Use the equations

$$t u_\nu'' + (2\nu + 1) u_\nu' + t u_\nu = 0, \qquad u_\nu(0) = \frac{1}{2^\nu \Gamma(\nu + 1)}$$

to prove that

$$\bar{u}_\nu(p) = (p^2 + 1)^{\nu - 1/2} \left( A_\nu - \frac{1}{2^{\nu - 1}\Gamma(\nu)} \int_0^p (z^2 + 1)^{-\nu - 1/2} \mathrm{d}z \right),$$

where $A_\nu$ is a constant. Show that if $\mathscr{R}\nu > 0$

$$\bar{u}_\nu(p) = \frac{(p^2+1)^{\nu-1/2}}{2^{\nu-1}\Gamma(\nu)} \int_p^\infty (z^2+1)^{-\nu-1/2}\,\mathrm{d}z,$$

and that

$$A_\nu = \frac{1}{2^\nu \Gamma(\nu+\frac{1}{2})}.$$

From the relation $u'_\nu(t) = -tu_{\nu+1}(t)$, prove that $A_\nu = (2\nu+1)A_{\nu+1}$, and hence that for all $\nu$

$$\bar{u}_\nu(p) = (p^2+1)^{\nu-1/2}\left(\frac{1}{2^\nu\Gamma(\nu+\frac{1}{2})} - \frac{1}{2^{\nu-1}\Gamma(\nu)}\int_0^p (z^2+1)^{-\nu-1/2}\,\mathrm{d}z\right).$$

**E2.5.20** Show that if $u(t)$ satisfies the equation

$$u'' - tu' - (a+\tfrac{1}{2})u = 0$$

and has a Laplace transform, then

$$\bar{u}(p) = \exp(-\tfrac{1}{2}p^2)p^{a-1/2}\left(\int_1^p \exp(\tfrac{1}{2}z^2)z^{-a-1/2}(zu_0+u_1)\,\mathrm{d}z + A\right),$$

where $u(0) = u_0$, $u'(0) = u_1$ and $A$ is a constant. Use the condition that $\bar{u}(p) \to 0$ as $|p| \to \infty$ with $|\arg p| \leqslant \frac{1}{2}\pi - \delta$ to prove that

$$u_1 = -\sqrt{2}\,\frac{\Gamma(\frac{1}{2}a+\frac{3}{4})}{\Gamma(\frac{1}{2}a+\frac{1}{4})}u_0.$$

[The solution with

$$u_0 = \frac{\sqrt{\pi}}{2^{a/2+1/4}\Gamma(\frac{1}{2}a+\frac{3}{4})}, \qquad u_1 = -\frac{\sqrt{\pi}}{2^{a/2-1/4}\Gamma(\frac{1}{2}a+\frac{1}{4})}$$

is $u(t) = \exp(\frac{1}{4}t^2)U(a,t)$, where $U(a,t) = D_{-a-1/2}(t)$ is a *parabolic cylinder function*.]

# 3
# Inverse Transforms

## 3.1 The Laplace Transform in the Complex Plane and Uniqueness

The inversion formula for the Laplace transformation requires the transform to be treated as a function of a complex variable $p$, whereas in the previous chapters $p$ could usually be thought of as real. We show first that the transform is a regular function of $p$.

**P3.1.1** If the Laplace transform integral

$$\bar{u}(p) = \int_0^\infty u(t)\mathrm{e}^{-pt}\,\mathrm{d}t$$

converges absolutely when $\mathscr{R}p = A$, then $\bar{u}(p)$ is a regular function of the complex variable $p$ in $\mathscr{R}p > A$ and

$$\frac{\mathrm{d}\bar{u}}{\mathrm{d}p} = -\int_0^\infty tu(t)\mathrm{e}^{-pt}\,\mathrm{d}t.$$

Let $p = x + \mathrm{i}y$. The integrals

$$\int_0^\infty t^n u(t)\mathrm{e}^{-pt}\,\mathrm{d}t \qquad (n = 0, 1, 2)$$

converge absolutely when $x > A$, and if $|h| < x - A$ then

$$\frac{\bar{u}(p+h) - \bar{u}(p)}{h} + \int_0^\infty tu(t)\mathrm{e}^{-pt}\,\mathrm{d}t = \int_0^\infty \frac{\mathrm{e}^{-ht} - 1 + ht}{h} u(t)\mathrm{e}^{-pt}\,\mathrm{d}t.$$

Also

$$|\mathrm{e}^{-ht} - 1 + ht| = \left|\sum_{n=2}^\infty \frac{(-ht)^n}{n!}\right| \leqslant \sum_{n=2}^\infty \frac{(|h|t)^n}{n!} = \mathrm{e}^{|h|t} - 1 - |h|t$$

for all $t \geqslant 0$, so that the modulus of the right-hand integral is at most

$$\int_0^\infty |u(t)|e^{-(x-|h|)t}[1-(1+|h|t)e^{-|h|t}]|h|^{-1}\,dt$$

$$\leqslant \int_0^\infty |u(t)|e^{-(x-|h|)t}|h|t^2\,dt,$$

since $e^{-|h|t} \geqslant 1-|h|t$. When $|h| \to 0$, we get the required result.

P3.1.1 provides the detailed justification of P1.1.9.

Although the integral defines $\bar{u}(p)$ only in a half plane, it may be possible to extend this definition into a larger region by analytic continuation, and Sections 3.3 and 3.4 will show how this may be used to obtain properties of $u(t)$ from those of $\bar{u}(p)$. Because $\bar{u}(p)$ is completely identified by its values in $\mathscr{R}p > A$, it is not usually necessary to state explicitly the region of convergence of a transform.

**E3.1.1** If

$$u(t) = \sum_{m=1}^{M} P_m(t)\exp(\alpha_m t)$$

is the exponential-type function of D1.1.2, its transform $\bar{u}(p)$, given in E1.1.7, has a pole at each $\alpha_m$ and the transform integral converges absolutely for $\mathscr{R}p > \max \mathscr{R}\alpha_m$.

The region of convergence need not be related to the singularities of the transform.

**E3.1.2** If $u(t) = \cos(t^2)$, the transform integral converges absolutely if $\mathscr{R}p > 0$ and simply if $\mathscr{R}p = 0$. The transform (E2.5.13)

$$\bar{u}(p) = \frac{1}{2}\int_p^\infty \sin\left[\tfrac{1}{4}(z^2-p^2)\right]dz$$

is regular for all values of $p$.

Before considering the inversion of the Laplace transformation, we should see how far a function is determined by its transform. If the value of $u(t)$ is changed at a single point $t_1$ the transform $\bar{u}(p)$ is unaltered, so we cannot expect $u(t)$ to be defined uniquely when $\bar{u}(p)$ is given unless we can impose on $u(t)$ some further condition such as continuity. We shall show that if $u_1(t)$ and $u_2(t)$ have the same transform, they are equivalent in the sense that they differ by a null function.

**D3.1.1** A *null function* $n(t)$ is defined by

$$\int_0^T n(t)\,\mathrm{d}t = 0 \qquad \text{for all } T \geqslant 0.$$

If $n(t)$ is continuous at $t = \tau$, then $n(\tau) = 0$.

**P3.1.2** If $n(t)$ is a null function, so is $\mathrm{e}^{\alpha t} n(t)$.

For

$$\int_0^T \mathrm{e}^{\alpha t} n(t)\,\mathrm{d}t = \left[ \mathrm{e}^{\alpha t} \int_0^t n(s)\,\mathrm{d}s \right]_0^T - \int_0^T \alpha \mathrm{e}^{\alpha t} \int_0^t n(s)\,\mathrm{d}s\,\mathrm{d}t,$$

and both terms are zero.

**P3.1.3** If $n(t)$ is a null function, then $\bar{n}(p) \equiv 0$.

For

$$\bar{n}(p) = \lim_{T \to \infty} \int_0^T n(t)\mathrm{e}^{-pt}\,\mathrm{d}t = 0$$

by P3.1.2.

**P3.1.4** *Uniqueness theorem.* If $\bar{u}_1(p) = \bar{u}_2(p)$ for $\mathscr{R} p > A$, then $u_1(t) - u_2(t)$ is a null function.

This follows from P3.1.5.

**P3.1.5** Suppose that

$$\int_0^\infty u(t)\mathrm{e}^{-\alpha t}\,\mathrm{d}t$$

is (simply) convergent, where $u(t)$ is absolutely integrable, and that $\bar{u}(\alpha + nk) = 0$ for some $k > 0$ and all positive integers $n$. Then $u(t)$ is a null function.

Note that the convergence of $\bar{u}(\alpha)$ implies the convergence of $\bar{u}(\alpha + nk)$, by E1.5.30 (ii). Let $0 < \tau < T$, and put $v(t) = u(t)\mathrm{e}^{-\alpha t}$. Then for any positive integer $n$

$$\begin{aligned} \int_0^T v(t)\mathrm{e}^{nk(T-t)}\,\mathrm{d}t &= -\int_T^\infty v(t)\mathrm{e}^{nk(T-t)}\,\mathrm{d}t \\ &= -\int_T^\infty \int_T^t v(s)\,\mathrm{d}s\; nk\mathrm{e}^{nk(T-t)}\,\mathrm{d}t, \end{aligned}$$

since $\int_T^\infty v(s)\,\mathrm{d}s$ converges. Also

$$\left| \int_T^t v(s)\,\mathrm{d}s \right| \leqslant M \qquad \text{(say) for all } t \geqslant T,$$

so that

$$\left|\int_0^T v(t)e^{nk(T-t)}\,dt\right| \leqslant M\int_T^\infty nk\,e^{nk(T-t)}\,dt = M.$$

The series

$$\sum_{m=1}^{\infty}\frac{(-1)^{m-1}}{m!}e^{mnk(\tau-t)}$$

converges uniformly in $0 \leqslant t \leqslant T$ to

$$E_n(\tau-t) = 1-\exp(-e^{nk(\tau-t)}).$$

Thus

$$\begin{aligned}\left|\int_0^T v(t)E_n(\tau-t)\,dt\right| &= \left|\sum_{m=1}^{\infty}\frac{(-1)^{m-1}}{m!}\int_0^T v(t)e^{mnk(\tau-t)}\,dt\right| \\ &\leqslant \sum_{m=1}^{\infty}\frac{1}{m!}e^{mnk(\tau-T)}\left|\int_0^T v(t)e^{mnk(\tau-T)}\,dt\right| \\ &\leqslant M[\exp(e^{-nk(T-\tau)})-1]\end{aligned}$$

and so tends to 0 as $n \to \infty$.

Now $E_n(\tau-t) \to H(\tau-t)$ as $n \to \infty$, and we can write

$$\int_0^T v(t)[E_n(\tau-t)-H(\tau-t)]\,dt = I_1+I_2+I_3+I_4,$$

where

$$I_1 = \int_0^{\tau-\delta_1} v(t)[E_n(\tau-t)-1]\,dt,$$

$$I_2 = \int_{\tau-\delta_1}^{\tau} v(t)[E_n(\tau-t)-1]\,dt,$$

$$I_3 = \int_{\tau}^{\tau+\delta_2} v(t)E_n(\tau-t)\,dt,$$

$$I_4 = \int_{\tau+\delta_2}^{T} v(t)E_n(\tau-t)\,dt.$$

But

$$|I_1| \leqslant N\exp[-\exp(nk\delta_1)],$$

$$|I_2| \leqslant \int_{\tau-\delta_1}^{\tau} |v(t)|\,dt,$$

$$|I_3| \leqslant \int_{\tau}^{\tau+\delta_2} |v(t)|\,\mathrm{d}t,$$

$$|I_4| \leqslant N\{1 - \exp[-\exp(-nk\delta_2)]\},$$

where

$$N = \int_0^T |v(t)|\,\mathrm{d}t.$$

We can therefore choose $\delta_1$ and $\delta_2$ to make $|I_2| < \varepsilon$ and $|I_3| < \varepsilon$, and then choose $n_0$ such that $|I_1| < \varepsilon$ and $|I_2| < \varepsilon$ for all $n \geqslant n_0$. Consequently

$$\int_0^{\tau} v(t)\,\mathrm{d}t = 0 \qquad \text{for any } \tau > 0.$$

Thus $v(t)$ is a null function and so, by P3.1.2, $u(t)$ is also null.

It follows from P3.1.4 that if $\bar{u}_1(p) \equiv \bar{u}_2(p)$ then $u_1(t) = u_2(t)$ for any value of $t$ at which the functions are continuous. It is natural to expect the inversion formula to pick a continuous function out of the class of equivalent functions wherever possible.

## 3.2 The Inversion Theorem

Before considering the general inversion formula it is instructive to look at two special classes of functions, the exponential-type functions and the step functions.

**P3.2.1** If $u(t)$ is an e.t.f., it is equal to the sum of the residues of $\bar{u}(p)\mathrm{e}^{pt}$. A typical term of $u(t)$ is $v(t) = at^m \mathrm{e}^{\alpha t}$, for which

$$\bar{v}(p) = m!\,a(p-\alpha)^{-m-1}.$$

Then

$$v(p)\mathrm{e}^{pt} = m!\,a\mathrm{e}^{\alpha t} \sum_{r=0}^{\infty} t^r (p-\alpha)^{r-m-1}/r!,$$

and the coefficient of $(p-\alpha)^{-1}$ is $m!\,a\mathrm{e}^{\alpha t}(t^m/m!) = v(t)$. Since $u(t)$ is the sum of terms such as $v(t)$, the general result follows.

From P3.2.1,

$$u(t) = \frac{1}{2\pi \mathrm{i}} \int_C \bar{u}(p)\mathrm{e}^{pt}\,\mathrm{d}p,$$

where $C$ is any simple closed contour in the $p$ plane containing all

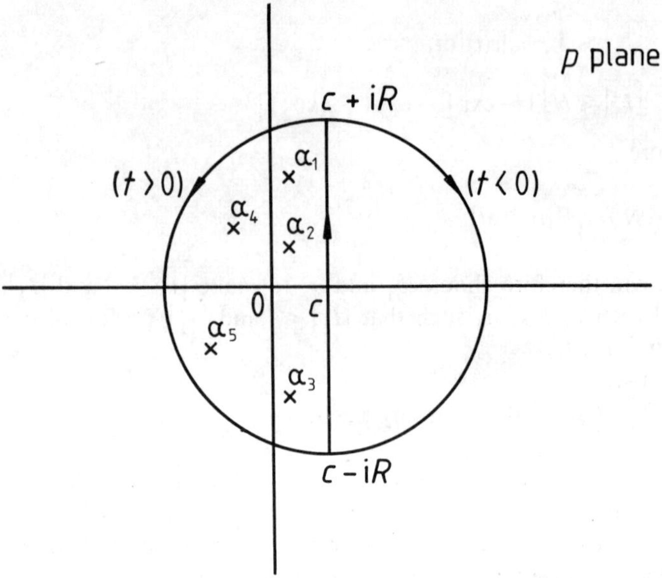


Figure 3.1

the poles of $\bar{u}(p)$. We can modify the integral to give zero when $t<0$ and $u(t)$ when $t>0$ by replacing $C$ by a straight line parallel to the imaginary axis of the $p$ plane; thus

$$u(t)\,\mathrm{H}(t) = \lim_{R\to\infty} \frac{1}{2\pi\mathrm{i}} \int_{c-\mathrm{i}R}^{c+\mathrm{i}R} \bar{u}(p)\mathrm{e}^{pt}\,\mathrm{d}p,$$

where all the poles of $\bar{u}(p)$ lie in $\mathscr{R}p<c$ (see Figure 3.1). This is because we can apply Jordan's lemma to a large semicircle of radius $R$, on the right if $t<0$ and on the left if $t>0$ (see P3.3.1).

**P3.2.2** If $u(t)$ is the step function

$$u(t) = \sum_{r=1}^{N} k_r[\mathrm{H}(t-t_{r-1}) - \mathrm{H}(t-t_r)],$$

where $0 \leqslant t_0 < t_1 < t_2 < \ldots < t_N \leqslant \infty$, and $c>0$, then

$$\lim_{R\to\infty} \frac{1}{2\pi\mathrm{i}} \int_{c-\mathrm{i}R}^{c+\mathrm{i}R} \bar{u}(p)\mathrm{e}^{pt}\,\mathrm{d}p = \begin{cases} u(t) & \text{if } t \neq \text{any } t_r, \\ \tfrac{1}{2}[u(t_r+0)+u(t_r-0)] & \text{if } t = t_r. \end{cases}$$

The transform of the typical term $v(t) = k[\mathrm{H}(t-a) - \mathrm{H}(t-b)]$ is (E1.1.10)

$$\bar{v}(p) = k(\mathrm{e}^{-ap} - \mathrm{e}^{-bp})p^{-1}.$$

The function $p^{-1}e^{pt}$ is regular except for a simple pole at $p = 0$ with residue 1. Hence if $c > 0$

$$\lim_{R\to\infty}\int_{c-iR}^{c+iR} p^{-1}e^{pt}\,dp = \begin{cases} 0 & (t<0),\\ 2\pi i & (t>0), \end{cases}$$

by application of Jordan's lemma to the appropriate semicircle. For $t = 0$ we have

$$\int_{c-iR}^{c+iR} p^{-1}\,dp = \log\left(\frac{c+iR}{c-iR}\right) = 2i\tan^{-1}\left(\frac{R}{c}\right) \to \pi i \qquad \text{as } R\to\infty.$$

Thus, on replacing $t$ by $t - a$ and $t - b$, we obtain

$$\lim_{R\to\infty}\frac{1}{2\pi i}\int_{c-iR}^{c+iR}\bar{v}(p)e^{pt}\,dp = \begin{cases} 0 & (t<a \text{ or } t>b),\\ \tfrac{1}{2}k & (t=a \text{ or } t=b),\\ k & (a<t<b). \end{cases}$$

This proves the result for $v(t)$ and the general result for $u(t)$ follows by addition. If $t_N = \infty$, the term $H(t - t_N)$ is always 0.

Since any integrable function can be approximated by a step function, we expect the general result to be similar to P3.2.2. In particular, we can expect to obtain a value halfway between the two limits at a simple discontinuity, and at $t = 0$ to get $\frac{1}{2}u(0)$.

The inversion theorem is based on the Riemann–Lebesgue theorem.

**P3.2.3** If $f(t)$ is absolutely integrable in $a \leqslant t \leqslant b$ and $y$ is real, then

$$\int_a^b f(t)e^{iyt}\,dt \to 0 \qquad \text{as } y\to\pm\infty.$$

For any $\varepsilon > 0$ there exists a step function

$$\phi(t) = \sum_{n=1}^{N} k_n[H(t-t_{n-1}) - H(t-t_n)]$$

such that

$$\int_a^b |f(t)-\phi(t)|\,dt < \varepsilon.$$

Then

$$\left|\int_a^b [f(t)-\phi(t)]e^{iyt}\,dt\right| \leqslant \int_a^b |[f(t)-\phi(t)]e^{iyt}|\,dt < \varepsilon,$$

and

$$\int_a^b \phi(t)\mathrm{e}^{\mathrm{i}yt}\,\mathrm{d}t = \sum_{n=1}^{N} k_n \int_{t_{n-1}}^{t_n} \mathrm{e}^{\mathrm{i}yt}\,\mathrm{d}t$$

$$= \sum_{n=1}^{N} \frac{k_n}{\mathrm{i}y}[\exp(\mathrm{i}yt_n) - \exp(\mathrm{i}yt_{n-1})].$$

Hence

$$\left|\int_a^b f(t)\mathrm{e}^{\mathrm{i}yt}\,\mathrm{d}t\right| < \varepsilon + \frac{2}{|y|}\sum_{n=1}^{N} |k_n| < 2\varepsilon,$$

provided that

$$|y| > 2\varepsilon^{-1} \sum_{n=1}^{N} |k_n| .$$

The same method proves P3.2.4.

**P3.2.4** If $0 \leqslant a < b$ then

$$\int_a^b f(t)\mathrm{e}^{-(x+\mathrm{i}y)t}\,\mathrm{d}t \to 0$$

as $y \to \pm\infty$ uniformly in $x \geqslant 0$.

**P3.2.5** If $A$ is real and

$$\int_0^\infty |u(t)|\mathrm{e}^{-At}\,\mathrm{d}t$$

converges, then $\bar{u}(x+\mathrm{i}y) \to 0$ as $y \to \pm\infty$ uniformly in $x \geqslant A$.

Let $v(t) = u(t)\mathrm{e}^{-At}$. Then

$$\left|\int_T^\infty u(t)\mathrm{e}^{-(x+\mathrm{i}y)t}\,\mathrm{d}t\right| \leqslant \int_T^\infty |v(t)|\mathrm{e}^{-(x-A)t}\,\mathrm{d}t \leqslant \int_T^\infty |v(t)|\,\mathrm{d}t < \varepsilon$$

by choice of $T$, independently of $x$ and $y$. Also, from P3.2.4,

$$\int_0^T u(t)\mathrm{e}^{-(x+\mathrm{i}y)t}\,\mathrm{d}t = \int_0^T v(t)\mathrm{e}^{-(x-A+\mathrm{i}y)t}\,\mathrm{d}t \to 0$$

as $y \to \pm\infty$ uniformly in $x \geqslant A$.

Combining P1.2.8 and P3.2.5 we have P3.2.6.

**P3.2.6** If $\bar{u}(p)$ converges absolutely when $\mathscr{R}p = A$, then $\bar{u}(p) \to 0$

as $|p| \to \infty$ uniformly in $|\arg(p - A)| \leqslant \frac{1}{2}\pi$.

This result shows that $e^{-ap}$ is not a proper Laplace transform for any value of $a$.

The inversion theorem can be proved with a variety of conditions on $u(t)$. The following form is adequate for many applications.

**P3.2.7** Let

$$\bar{u}(p) = \int_0^\infty u(t) e^{-pt} \, dt$$

be absolutely convergent for $\mathscr{R}p = c$. Define $u(t) = 0$ for $t < 0$. Suppose that $u(\tau + 0)$ and $u(\tau - 0)$ exist such that as $s \to +0$

$$|u(\tau + s) - u(\tau + 0)| = O(s^a),$$
$$|u(\tau - s) - u(\tau - 0)| = O(s^a),$$

where $a > 0$. Then as $R \to \infty$

$$I(R) = \frac{1}{2\pi i} \int_{c - iR}^{c + iR} \bar{u}(p) e^{p\tau} \, dp \to \tfrac{1}{2}[u(\tau + 0) + u(\tau - 0)]$$

where the integration is along the straight line $\mathscr{R}p = c$.

Since $\bar{u}(p)$ is absolutely convergent for $p = c + iy$

$$I(R) = \frac{1}{2\pi i} \int_{-R}^{R} \bar{u}(c + iy) e^{(c + iy)\tau} \, i dy$$
$$= \frac{1}{2\pi} \int_0^\infty u(t) e^{c(\tau - t)} \int_{-R}^{R} e^{iy(\tau - t)} \, dy \, dt$$
$$= \frac{1}{2\pi} \int_0^\infty u(t) e^{c(\tau - t)} \frac{e^{iR(\tau - t)} - e^{-iR(\tau - t)}}{i(\tau - t)} \, dt$$
$$= \frac{1}{\pi} \int_{-\infty}^\infty u(\tau + s) e^{-cs} \frac{\sin Rs}{s} \, ds,$$

since $u(t) = 0$ for $t < 0$. Now let

$$I_+(R) = \frac{1}{\pi} \int_0^\infty u(\tau + s) e^{-cs} \frac{\sin Rs}{s} \, ds.$$

Then because (E1.5.31)

$$\int_0^\infty s^{-1} \sin Rs \, ds = \tfrac{1}{2}\pi \qquad \text{for } R > 0,$$

we have

$$\pi I_+(R) - \tfrac{1}{2}\pi u(\tau+0) = \int_0^\infty [u(\tau+s)e^{-cs} - u(\tau+0)]\frac{\sin Rs}{s}\,ds$$

$$= I_1 + I_2 + I_3,$$

where

$$I_1 = \int_0^X s^{-1}[u(\tau+s)e^{-cs} - u(\tau+0)]\sin Rs\,ds,$$

$$I_2 = \int_X^\infty u(\tau+s)e^{-cs}\frac{\sin Rs}{s}\,ds,$$

$$I_3 = -\int_X^\infty u(\tau+0)\frac{\sin Rs}{s}\,ds.$$

As $X \to \infty$ both $I_2$ and $I_3 \to 0$, so we can choose $X$ so great that $|I_2 + I_3| < \varepsilon$. Also

$$f_+(s) = s^{-1}[u(\tau+s)e^{-cs} - u(\tau+0)] = O(s^{a-1})$$

as $s \to 0$, so that $f_+(s)$ is absolutely integrable in $0 \leqslant s \leqslant X$. From P3.2.3 $I_1 \to 0$ as $R \to \infty$, so that $|I_1| < \varepsilon$ for $R > R_0(\varepsilon)$ and therefore

$$|I_+(R) - \tfrac{1}{2}u(\tau+0)| < (2/\pi)\varepsilon.$$

Hence, $I_+(R) \to \tfrac{1}{2}u(\tau+0)$ as $R \to \infty$ and similarly

$$I_-(R) = \frac{1}{\pi}\int_{-\infty}^0 u(\tau+s)e^{-cs}\frac{\sin Rs}{s}\,ds \to \tfrac{1}{2}u(\tau-0).$$

Thus

$$\lim_{R\to\infty} I(R) = \tfrac{1}{2}[u(\tau+0) + u(\tau-0)].$$

Note that the conditions of P3.2.7 are satisfied with $a = 1$ when the function $u(t)$ has right and left derivatives at $t = \tau$, that is

$$u(\tau+s) = u(\tau+0) + u'(\tau+0)s + O(s),$$
$$u(\tau-s) = u(\tau-0) - u'(\tau-0)s + O(s).$$

P3.2.7 is a theorem on the convergence of the inverse transform at the particular point $t = \tau$. If the conditions hold for all $t$, or throughout some interval, then the inversion formula applies for all $t$, or throughout the interval. In particular, if $u(t)$ is piecewise differentiable, so that $u(t \pm 0)$, $u'(t \pm 0)$ exist for all $t$, the inverse transform gives $\frac{1}{2}[u(t+0) + u(t-0)]$ for all $t$.

The inverse transform will usually be written as

$$u(t) = \mathscr{L}^{-1}\{\bar{u}(p)\} = \frac{1}{2\pi i} \int_{c-i\infty}^{c+i\infty} \bar{u}(p) e^{pt} \, dp,$$

but it must be remembered that the meaning of this is as stated in P3.2.7. We could start with $\bar{u}(p)$ as an arbitrary regular function such that (P3.2.6) $\bar{u}(p) \to 0$ uniformly as $|p| \to \infty$ with $|\arg(p - A)| \leqslant \frac{1}{2}\pi$. However, the inverse transform $u(t)$ need not be integrable in the sense of Riemann, so that a more general theory of integration, such as that of Lebesgue, is required in this approach.

## 3.3 Evaluation of Inverse Transforms

As noted in Section 3.1, the transform integral defines $\bar{u}(p)$ as a regular function of $p$ in the half plane of absolute convergence, but the range of definition may be extended by analytic continuation. If no natural boundary is met, the function $\bar{u}(p)$ may be regular for all $p$ or it may have poles, branch points or essential singularities of some kind. It is convenient to introduce a cut in the $p$ plane from any branch point to $-\infty$, usually along a line parallel to the real axis. The inverse transform can often be expressed as a sum of integrals round the poles and loop integrals round the cuts from the branch points. For this purpose, the following general form of Jordan's lemma is useful.

**P3.3.1** Let $C(r)$ be a path in the complex plane of $z = x + iy$, lying wholly in the half plane $x \geqslant A$ and depending on the positive parameter $r$. Suppose that the length of $C(r)$ is at most $kr$ and that $C(r)$ consists of two parts $C_1(r)$ and $C_2(r)$ such that

(i) on $C_1(r)$, $\quad |dz| \leqslant b\, dx$

(ii) on $C_2(r)$, $\quad x \geqslant cr^a$

where $a, b, c$ and $k$ are positive constants. Then, if $s > 0$ and $|f(z)| \leqslant M(r)$ for $z$ on $C(r)$, where $M(r) \to 0$ as $r \to \infty$ (taking either discrete or continuous values),

$$\int_{C(r)} f(z) e^{-sz} \, dz \to 0 \qquad \text{as } r \to \infty.$$

For

$$\left|\int_{C_1(r)}\right| \leqslant \int_A^\infty M(r)\mathrm{e}^{-sx} b\,\mathrm{d}x,$$

$$\left|\int_{C_2(r)}\right| \leqslant M(r)\exp(-scr^a)kr.$$

In an application to Laplace transforms, we usually have $s=|t|$ and $z=-(\operatorname{sgn} t)p$; $C(r)$ is often a quadrant of a circle, since two quadrants form the semicircle of Jordan's lemma, but sometimes (see E3.3.2, E4.2.2 and E4.3.3) it is convenient to determine $M(r)$ on a parabola or on a straight line in the $p$ plane, and then $C(r)$ is chosen accordingly.

If $\bar{u}(p)$ is a rational function of $p$ and $\mathrm{O}(p^{-1})$ at infinity, then $u(t)$ is an e.t.f. As shown in P1.1.8, we can find $u(t)$ by expressing $\bar{u}(p)$ in partial fractions. When there are repeated factors, it is usually easier to apply the residue method of P3.2.1.

**E3.3.1**

$$\bar{u}(p)=\frac{p^2-k^2}{p(p^2+k^2)^3}$$

has a simple pole at $p=0$ and triple poles at $p=\pm\mathrm{i}k$. The residue of $\bar{u}(p)\mathrm{e}^{pt}$ at $p=0$ is $-k^{-4}$. To find the residue at $p=\mathrm{i}k$, set $p=\mathrm{i}k+\delta$. Then

$$\begin{aligned}
\bar{u}(p)\mathrm{e}^{pt} &= \frac{(\mathrm{i}k+\delta)^2-k^2}{(\mathrm{i}k+\delta)\delta^3(2\mathrm{i}k+\delta)^3}\mathrm{e}^{(\mathrm{i}k+\delta)t} \\
&= \frac{-2k^2+2\mathrm{i}k\delta+\delta^2}{\mathrm{i}k\delta^3(2\mathrm{i}k)^3}\left(1+\frac{\delta}{\mathrm{i}k}\right)^{-1}\left(1+\frac{\delta}{2\mathrm{i}k}\right)^{-3}\mathrm{e}^{(\mathrm{i}k+\delta)t} \\
&= \frac{\mathrm{e}^{\mathrm{i}kt}}{8k^4\delta^3}(-2k^2+2\mathrm{i}k\delta+\delta^2)\left(1-\frac{\delta}{\mathrm{i}k}-\frac{\delta^2}{k^2}\right) \\
&\quad\times\left(1-\frac{3\delta}{2\mathrm{i}k}-\frac{3\delta^2}{2k^2}\right)(1+\delta t+\tfrac{1}{2}\delta^2t^2)+\mathrm{O}(1),
\end{aligned}$$

on expanding in powers of $\delta$. The coefficient of $\delta^{-1}$ is

$$\frac{\mathrm{e}^{\mathrm{i}kt}}{8k^4}(-k^2t^2-3\mathrm{i}kt+4).$$

Similarly, the residue at $p = -ik$ is

$$\frac{e^{-ikt}}{8k^4}(-k^2t^2 + 3ikt + 4)$$

and therefore

$$u(t) = k^{-4}[-1 + (1 - \tfrac{1}{4}k^2t^2)\cos kt + \tfrac{3}{4}kt \sin kt].$$

If the transform $\bar{u}(p)$ has an infinite number of poles, with their only limit point at infinity, it may be possible to enclose increasing numbers of poles between the line $\mathscr{R}p = c$ and a sequence of curves $C_N$ such that

$$\int_{C_N} \bar{u}(p)e^{pt}\,dp \to 0 \qquad \text{as } N \to \infty.$$

In this case, $u(t)$ is the infinite sum of the residues at the poles. In this way the Fourier series of a periodic function may be obtained from its Laplace transform (Section 5.3). Another example is the following.

**E3.3.2** $\bar{u}(p) = p^{-1/2} \coth p^{1/2}$ does not have a branch point at the origin since each value of the square root gives the same value of $\bar{u}(p)$. The singularities of $\bar{u}(p)$ are simple poles at the origin and at $p = -n^2\pi^2$ for $n = 1, 2, 3, \ldots$

On the parabola $C_N$ (see Figure 3.2) given by $p = [(N + \tfrac{1}{2})\pi i + \xi]^2$,

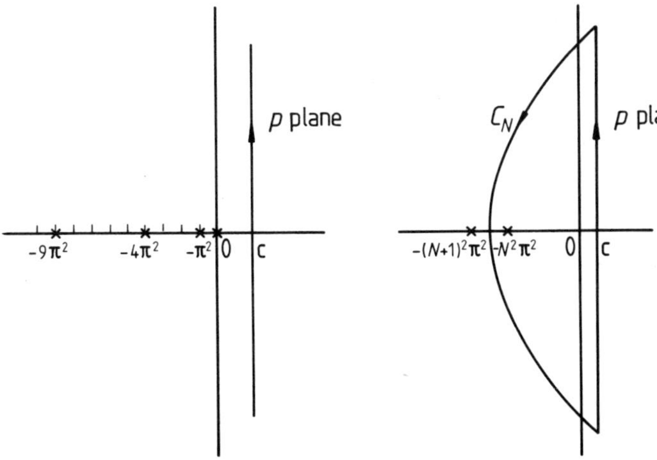


Figure 3.2

where $\xi$ is real, $\coth p^{1/2} = \tanh \xi$ so that

$$|\bar{u}(p)| \leqslant \frac{1}{(N+\frac{1}{2})\pi}.$$

P3.3.1 now shows that if $t>0$ the integral along the part of $C_N$ with $\mathscr{R}p<c$ tends to zero as $N \to \infty$. Thus $u(t)$ is the sum of the residues of $\bar{u}(p)e^{pt}$. The residue at $p=0$ is 1, and the residue at $p=-n^2\pi^2$ is $2\exp(-n^2\pi^2 t)$, so that

$$\bar{u}(t) = 1 + 2\sum_{n=1}^{\infty} \exp(-n^2\pi^2 t) \qquad \text{for all } t>0.$$

The simplest example of a transform with a branch point is provided by E1.2.2.

**E3.3.3** Since $p^{-\nu}$ is regular in $\mathscr{R}p>0$ and tends to zero as $|p| \to \infty$ when $\mathscr{R}\nu > 0$ we have

$$\mathscr{L}^{-1}\{p^{-\nu}\} = \frac{1}{2\pi i}\int_{c-i\infty}^{c+i\infty} p^{-\nu}e^{pt}\,dp$$

for $c>0$. From E1.2.2, an application of Jordan's lemma gives

$$\frac{t^{\nu-1}}{\Gamma(\nu)} = \frac{1}{2\pi i}\int_{-\infty}^{(0+)} p^{-\nu}e^{pt}\,dp \qquad \text{for } t>0. \tag{3.3.1}$$

The path of integration can be taken to coincide with the edges of the cut along the negative real axis together with a circle of radius $\delta$ (see Figure 3.3).

If also $\mathscr{R}\nu < 1$, the integral round this circle tends to zero as $\delta \to 0$, and then the substitutions $p = re^{\pm\pi i}$ on the two sides of the

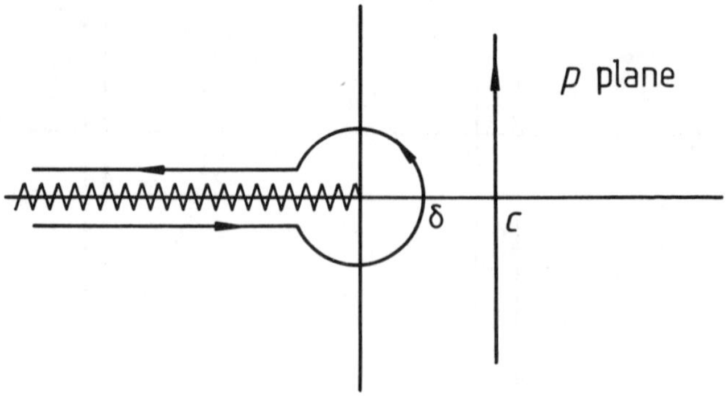


Figure 2.6

cut give

$$\frac{t^{\nu-1}}{\Gamma(\nu)} = \frac{1}{2\pi\mathrm{i}}\int_{\infty}^{0} r^{-\nu}\mathrm{e}^{\nu\pi\mathrm{i}}\mathrm{e}^{-rt}(-\mathrm{d}r) + \frac{1}{2\pi\mathrm{i}}\int_{0}^{\infty} r^{-\nu}\mathrm{e}^{-\nu\pi\mathrm{i}}\mathrm{e}^{-rt}(-\mathrm{d}r)$$

$$= \frac{\mathrm{e}^{\nu\pi\mathrm{i}} - \mathrm{e}^{-\nu\pi\mathrm{i}}}{2\pi\mathrm{i}}\int_{0}^{\infty} r^{-\nu}\mathrm{e}^{-rt}\,\mathrm{d}r$$

$$= \frac{\sin\nu\pi}{\pi}\,\frac{\Gamma(1-\nu)}{t^{1-\nu}},$$

by E1.2.2 with $t$ for $p$ and $r$ for $t$. Thus the inversion theorem gives the functional equation

$$\Gamma(\nu)\Gamma(1-\nu) = \pi \operatorname{cosec} \nu\pi \qquad (3.3.2)$$

for $0 < \mathscr{R}\nu < 1$. Analytic continuation shows that the loop integral (3.3.1) is valid for all $\nu$, and the functional equation (3.3.2) for all non-integral $\nu$.

The method of series expansion for the inverse transform has already been given as P1.4.1, but it can also be derived from the inversion integral as follows.

**P3.3.2** If

$$\bar{u}(p) = \sum_{n=0}^{\infty} a_n p^{-n-\nu},$$

where $\mathscr{R}\nu > 0$ and the series converges for $|p| \geqslant R$, then

$$u(t) = \sum_{n=0}^{\infty} a_n \frac{t^{n+\nu-1}}{\Gamma(n+\nu)},$$

where the series converges for all $t > 0$.

The path of integration in P3.2.7 can be deformed so as to lie in $|p| > R$, $\mathscr{R}p \geqslant c$, and in this region the series for $\bar{u}(p)$ is uniformly convergent.

Another expansion procedure, which will be useful in Chapter 4, is illustrated in the next two examples.

**E3.3.4** If $a > 0$, the function $\bar{u}(p) = [p(1-\mathrm{e}^{-ap})]^{-1}$ is regular in $\mathscr{R}p > 0$ and we can write

$$\bar{u}(p) = \sum_{n=0}^{\infty} p^{-1}\mathrm{e}^{-nap},$$

a series which converges uniformly in $\mathscr{R}p \geqslant A > 0$. Hence

$$u(t) = \sum_{n=0}^{\infty} \mathrm{H}(t - na) = m + 1 \qquad \text{when } ma < t < (m+1)a.$$

**E3.3.5**

$$\begin{aligned}\bar{u}(p) &= p^{-1/2} \coth(\sqrt{p}) \\ &= \frac{1}{\sqrt{p}} \frac{1 + \exp(-2\sqrt{p})}{1 - \exp(-2\sqrt{p})} \\ &= \frac{1}{\sqrt{p}} \left(1 + 2 \sum_{n=1}^{\infty} \exp(-2n\sqrt{p})\right).\end{aligned}$$

The series converges uniformly when $\mathscr{R}p^{1/2} \geqslant A > 0$, that is except in a parabola enclosing the negative real axis. From E1.3.10 we have

$$u(t) = (\pi t)^{-1/2} \left[1 + 2 \sum_{n=1}^{\infty} \exp\left(-\frac{n^2}{t}\right)\right] \qquad \text{for } t > 0.$$

This series converges very rapidly unless $t$ is large; the series obtained in E3.3.2 converges very rapidly unless $t$ is quite small.

It is sometimes useful to obtain a differential equation for the inverse transform. The following example will be useful in Chapter 4.

**E3.3.6** $\bar{u}_k(x, p) = p^{-1-k/2} \exp(-x\sqrt{p})$ is a Laplace transform if $|\arg x| < \frac{1}{4}\pi$, and also when $\arg x = \pm\frac{1}{4}\pi$ provided that $\mathscr{R}k > -2$.

When $t > 0$, the inverse transform is (compare E3.3.3)

$$u_k(x, t) = \frac{1}{2\pi \mathrm{i}} \int_{-\infty}^{(0+)} p^{-1-k/2} \exp(pt - x\sqrt{p}) \mathrm{d}p \tag{3.3.3}$$

by Jordan's lemma. This can be expressed in a standard form by the substitution $p = (\eta + \mathrm{i}\xi)^2/2t$, where $\eta = x/\sqrt{(2t)}$. Thus

$$u_k(x, t) = \sqrt{(2/\pi)}(2t)^{k/2} g_k(\eta)$$

where

$$g_k(\eta) = (2\pi)^{-1/2} \int_{-\infty}^{\infty} (\eta + \mathrm{i}\xi)^{-1-k} \exp(-\tfrac{1}{2}\eta^2 - \tfrac{1}{2}\xi^2) \mathrm{d}\xi. \tag{3.3.4}$$

If we divide the range of integration at $\xi = 0$, we can take $\xi^2/\eta^2$ as the new variable of integration and apply P1.2.13 to show that

$$g_k(\eta) \sim \eta^{-1-k} \exp(-\tfrac{1}{2}\eta^2) \tag{3.3.5}$$

as $\eta \to \infty$ with $|\arg \eta| < \frac{1}{4}\pi$. Thus $u_k(x, t)$ is exponentially small as $t \to 0$ if $x > 0$ or as $x \to \infty$ when $t > 0$.

The power series for $g_k(\eta)$ is obtained by expanding $\exp(-x\sqrt{p})$ in (3.3.3). Thus

$$u_k(x, t) = \sum_{n=0}^{\infty} \frac{(-x)^n}{n!} \frac{1}{2\pi i} \int_{-\infty}^{(0+)} p^{(n-k)/2-1} e^{pt} dp$$

$$= \sum_{n=0}^{\infty} \frac{(-x)^n t^{(k-n)/2}}{n! \Gamma(\frac{1}{2}k - \frac{1}{2}n + 1)},$$

so that

$$g_k(\eta) = (\tfrac{1}{2}\pi) \sum_{n=0}^{\infty} \frac{2^{(n-k)/2}(-\eta)^n}{n! \Gamma(\frac{1}{2}k - \frac{1}{2}n + 1)}$$

$$= \sqrt{(\tfrac{1}{2}\pi)} \left( \frac{2^{-k/2}}{\Gamma(\frac{1}{2}k + 1)} {}_1F_1(-\tfrac{1}{2}k; \tfrac{1}{2}; -\tfrac{1}{2}\eta^2) - \frac{2^{(1-k)/2}}{\Gamma(\frac{1}{2}k + \frac{1}{2})} {}_1F_1(\tfrac{1}{2} - \tfrac{1}{2}k; \tfrac{3}{2}; -\tfrac{1}{2}\eta^2) \right)$$

$$= 2^{-(1+k)/2} \exp(-\tfrac{1}{2}\eta^2) U(\tfrac{1}{2} + \tfrac{1}{2}k; \tfrac{1}{2}; \tfrac{1}{2}\eta^2),$$

in the notation of confluent hypergeometric functions (see D1.5.3 and E1.5.19 (i)).

We can also get a differential equation and recurrence relations for $g_k(\eta)$. From $\partial \bar{u}_k / \partial x = -\bar{u}_{k-1}$, $p\bar{u}_k = \bar{u}_{k-2}$, we have

$$\partial u_k / \partial x = -u_{k-1}, \qquad \partial u_k / \partial t = u_{k-2} = \partial^2 u_k / \partial x^2.$$

In terms of $\eta$, these equations give

$$g_k'(\eta) = -g_{k-1}(\eta) \tag{3.3.6}$$

and

$$g_k''(\eta) + \eta g_k'(\eta) - k g_k(\eta) = 0, \tag{3.3.7}$$

so that

$$g_{k-2}(\eta) - \eta g_{k-1}(\eta) - k g_k(\eta) = 0. \tag{3.3.8}$$

Equation (3.3.7) is related to the equation of parabolic cylinder functions

$$G_k''(\eta) - (\tfrac{1}{4}\eta^2 + k + \tfrac{1}{2}) G_k(\eta) = 0$$

by the substitution $g_k(\eta) = \exp(-\frac{1}{4}\eta^2) G_k(\eta)$. The asymptotic form

(3.3.5) shows that, in the notations for these functions,

$$g_k(\eta) = \exp(-\tfrac{1}{4}\eta^2)D_{-k-1}(\eta) = \exp(-\tfrac{1}{4}\eta^2)U(k+\tfrac{1}{2}, \eta).$$

When $2k$ is an integer, $g_k(\eta)$ can be expressed in terms of more familiar functions. From (3.3.4),

$$g_{-1}(\eta) = \exp(-\tfrac{1}{2}\eta^2),$$

and so from (3.3.6)

$$g_{-n-1}(\eta) = (-1)^n \frac{\mathrm{d}^n}{\mathrm{d}\eta^n}(\exp(-\tfrac{1}{2}\eta^2)) = h_n(\eta)\exp(-\tfrac{1}{2}\eta^2),$$

where $h_n(\eta)$ is a Hermite polynomial. The cases $n = 0, 1$ were given in E1.3.10. By integration and use of equation (3.3.5), we have

$$g_0(\eta) = \sqrt{(\tfrac{1}{2}\pi)}\,\mathrm{erfc}(\eta/\sqrt{2}),$$

and hence

$$g_1(\eta) = \sqrt{\pi}\,\mathrm{ierfc}(\eta/\sqrt{2})$$

and

$$g_n(\eta) = 2^{(n-1)/2}\sqrt{\pi}\,\mathrm{i}^n\,\mathrm{erfc}(\eta/\sqrt{2}),$$

where

$$\mathrm{i}^n\,\mathrm{erfc}(\theta) = \int_\theta^\infty \mathrm{i}^{n-1}\,\mathrm{erfc}(\phi)\,\mathrm{d}\phi, \qquad \mathrm{i}^0\,\mathrm{erfc}(\theta) = \mathrm{erfc}(\theta).$$

The recurrence relation (3.3.8) enables $g_n(\eta)$ to be expressed in terms of $g_0(\eta)$ and $g_{-1}(\eta)$.

The equation for $G_{-1/2}(\eta)$ has solutions $\eta^{1/2}\,\mathrm{I}_{\pm 1/4}(\tfrac{1}{4}\eta^2)$, and the condition (3.3.5) shows that

$$g_{-1/2}(\eta) = \left(\frac{\eta}{2\pi}\right)^{1/2} \exp(-\tfrac{1}{4}\eta^2)\mathrm{K}_{1/4}(\tfrac{1}{4}\eta^2).$$

Hence

$$g_{-3/2}(\eta) = -g'_{-1/2}(\eta) = \pi^{-1/2}(\tfrac{1}{2}\eta)^{3/2}\exp(-\tfrac{1}{4}\eta^2) \times [\mathrm{K}_{1/4}(\tfrac{1}{4}\eta^2) + \mathrm{K}_{3/4}(\tfrac{1}{4}\eta^2)],$$

and other functions $g_{n+1/2}(\eta)$ can be expressed in terms of $\mathrm{K}_{1/4}(\tfrac{1}{4}\eta^2)$ and $\mathrm{K}_{3/4}(\tfrac{1}{4}\eta^2)$ by means of the recurrence relation (3.3.8).

## 3.4 Asymptotic Expansion of Inverse Transforms

The solution of a problem by the Laplace transform method may produce a transform that cannot be inverted in terms of standard functions, and the expansion procedures of Section 3.3 may not give convergent series for the inverse transform. It is therefore useful to be able to find asymptotic formulae for the transform when $t$ is large or small.

**P3.4.1** Suppose that $\bar{u}(p)$ is regular in $\mathscr{R}p \geqslant A > 0$ and that

$$\bar{u}(p) \sim \sum_{n=0}^{\infty} a_n p^{-n-\nu},$$

where $\mathscr{R}\nu > 0$, as $p \to \infty$ uniformly in $|\arg(p-A)| \leqslant \frac{1}{2}\pi$. Then as $t \to 0+$

$$u(t) \sim \sum_{n=0}^{\infty} \frac{a_n}{\Gamma(n+\nu)} t^{n+\nu-1}.$$

We can write

$$\bar{u}(p) = \sum_{n=0}^{N-1} a_n p^{-n-\nu} + \bar{u}_N(p),$$

where $|\bar{u}_N(p)| \leqslant K|p^{-N-\nu}|$ for $\mathscr{R}p \geqslant A$. Then

$$u(t) = \sum_{n=0}^{N-1} \frac{a_n}{\Gamma(n+\nu)} t^{n+\nu-1} + u_N(t),$$

where

$$u_N(t) = \frac{1}{2\pi \mathrm{i}} \int_{c-\mathrm{i}\infty}^{c+\mathrm{i}\infty} \bar{u}_N(p) \mathrm{e}^{pt} \,\mathrm{d}p$$

for any $c > A$.

For any $t$ in the range $0 < t < 1$, choose $c = A/t$, and put

$$p = \frac{A}{t}(1 + \mathrm{i}\eta).$$

Then if $\nu = \nu_1 + \mathrm{i}\nu_2$

$$|\bar{u}_N(p)| \leqslant K\left|\left(\frac{A}{t}(1+\mathrm{i}\eta)\right)^{-N-\nu}\right|$$

$$\leqslant K\left(\frac{t}{A}\right)^{N+\nu_1}(1+\eta^2)^{-(N+\nu_1)/2}\exp(\tfrac{1}{2}\pi|\nu_2|),$$

so that

$$|u_N(t)| \leqslant \frac{1}{2\pi}\int_{-\infty}^{\infty} K\left(\frac{t}{A}\right)^{N+\nu_1-1} (1+\eta^2)^{-(N+\nu_1)/2}$$
$$\times \exp(\tfrac{1}{2}\pi|\nu_2| + A)\,d\eta$$
$$\leqslant Lt^{N+\nu_1-1},$$

where $L$ is independent of $t$.

Note that, although the converse P1.2.13 gave the asymptotic expansion of $\bar{u}(p)$ only for $|\arg p| \leqslant \frac{1}{2}\pi - \delta$, here we assume that it holds as $\arg p \to \pm\frac{1}{2}\pi$.

The argument is easily extended to include terms involving $\log p$. The form of the result is obtained by differentiating with respect to $\nu$.

**P3.4.2** Suppose that $\bar{v}(p)$ is regular in $\mathscr{R}p \geqslant A > 0$ and that

$$\bar{v}(p) \sim \sum_{n=0}^{\infty} b_n p^{-n-\nu} \log p,$$

where $\mathscr{R}\nu > 0$, as $p \to \infty$ uniformly in $|\arg(p-A)| \leqslant \frac{1}{2}\pi$. Then as $t \to 0+$

$$v(t) \sim \sum_{n=0}^{\infty} \frac{b_n}{\Gamma(n+\nu)} t^{n+\nu-1} [\psi(n+\nu) - \log t].$$

If

$$\bar{v}_N(p) = \bar{v}(p) - \sum_{n=0}^{N-1} b_n p^{-n-\nu} \log p$$

then $|\bar{v}_N(p)| \leqslant K|p^{-N+1/2-\nu}|$ in $\mathscr{R}p \geqslant A$.

**E3.4.1** $\bar{u}(p) = p^{-1}\psi(p)$. From E1.5.38 (i) we have

$$\bar{u}(p) \sim p^{-1}\log p - \tfrac{1}{2}p^{-2} + \sum_{n=1}^{\infty} \frac{(-1)^n B_n}{2np^{2n+1}},$$

where the $B_n$ are Bernoulli numbers. Hence as $t \to 0+$

$$u(t) \sim -(\log t + \gamma) - \tfrac{1}{2}t + \sum_{n=1}^{\infty} \frac{(-1)^n B_n}{2n} \frac{t^{2n}}{(2n)!}.$$

The inverse transform can be evaluated exactly by the residue method. $\psi(p)$ has simple poles at $p = 0, -1, -2, \ldots$ with residue $-1$, and near $p = 0$

$$\bar{u}(p) = p^{-1}[\psi(p+1) - p^{-1}] = -p^2 - \gamma p^{-1} + \mathrm{O}(1).$$

Hence

$$u(t) = -t - \gamma + \sum_{n=1}^{\infty} n^{-1}\mathrm{e}^{-nt} = -t - \gamma - \log(1 - \mathrm{e}^{-t}).$$

In this example, the asymptotic expansion converges to the exact value of $u(t)$, but this is not true in general. If

$$v(t) = t^{-1/2}\exp(-a/t), \qquad \text{where } a > 0,$$

then from E1.3.10

$$\bar{v}(p) = \sqrt{(\pi/p)}\exp[-2\sqrt{(ap)}].$$

Since $\bar{v}(p) = \mathrm{o}(p^{-n})$ for any $n$ as $p \to \infty$ in $\mathscr{R}p \geqslant 0$, we could add any multiple of $\bar{v}(p)$ to $\bar{u}(p)$ without affecting the asymptotic series, and since $v(t) = \mathrm{o}(t^n)$ for any $n$ as $t \to 0+$, the corresponding addition to $u(t)$ is undetectable in its asymptotic expansion.

To find the asymptotic behaviour of an inverse transform as $t \to \infty$, we move the path of integration to the left. Suppose that $\bar{u}(p)$ can be defined by analytic continuation in a half plane $\mathscr{R}p > B$ with the exception of a finite number of poles and branch points, and that the path of integration can be moved to the left from the straight line $\mathscr{R}p = c$ to give the line $\mathscr{R}p = b > B$ together with integrals round the poles $\alpha_m$ and the branch points $\beta_m$ in $\mathscr{R}p > b$ (see Figure 3.4). Suppose also that $\bar{u}(p)$ is regular on $\mathscr{R}p = b$.

The integral round a pole of order $N$ at $p = \alpha$ gives a contribution

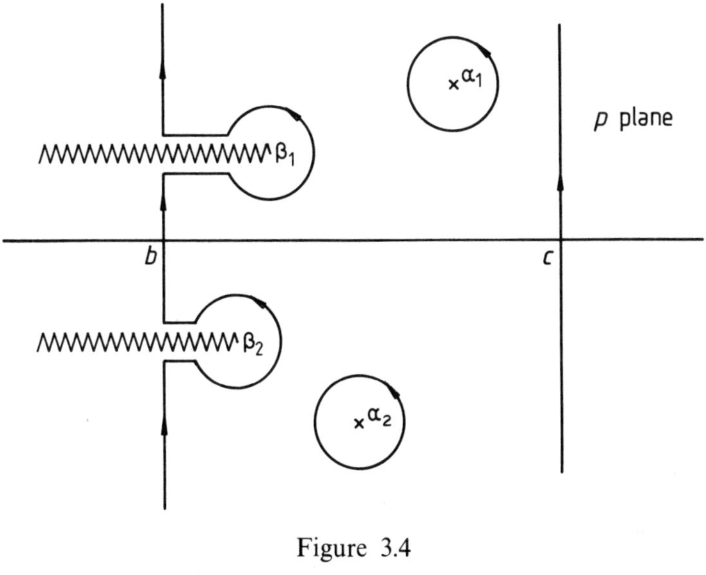


Figure 3.4

to $u(t)$ of the form

$$u(t;\alpha) = \frac{1}{2\pi i}\int^{(\alpha+)} \sum_{n=1}^{N} a_n(p-\alpha)^{-n} e^{pt}\,dp = \sum_{n=1}^{N} a_n \frac{t^{n-1}}{(n-1)!} e^{\alpha t},$$

where

$$\sum_{n=1}^{N} a_n(p-\alpha)^{-n}$$

is the principal part of $\bar{u}(p)$ at $p = \alpha$.

There is an analogous result for a branch point, but the contribution is, in general, an infinite series which represents an asymptotic expansion for $t \to \infty$. Suppose that when $|p - \beta| < r$ the transform can be expanded as

$$\bar{u}(p) = \sum_{n=0}^{\infty} b_n(p-\beta)^{n+\nu}.$$

Then the loop integral round the branch point $\beta$ gives the formal asymptotic series

$$u(t;\beta) \sim \sum_{n=0}^{\infty} \frac{b_n}{\Gamma(-n-\nu)} t^{-n-\nu-1} e^{\beta t}$$

$$\sim \frac{\sin \nu\pi}{\pi} \sum_{n=0}^{\infty} (-1)^{n+1} b_n \Gamma(n+\nu+1) t^{-n-\nu-1} e^{\beta t},$$

from equations (3.3.1) and (3.3.2), as will be shown in P3.4.5.

On $\mathscr{R}p = b$ the integrand is $O(e^{bt})$, so that we expect this part of the integral to be negligible when $t$ is large, compared with the contributions from the singularities. For large $t$ the most important parts of the inverse transform come from the singularities with greatest real part.

**P3.4.3** Let $\bar{u}(p)$ be regular for $\mathscr{R}p > b$ except for a finite number of poles $\alpha_m$ and branch points $\beta_m$, when the plane is cut from each branch point to $-\infty$, and let $\bar{u}(p)$ be regular on $\mathscr{R}p = b$. Suppose that $\bar{u}(x + iy)$ and $\bar{u}(x - iy)$ both tend uniformly to zero in $b \leqslant x \leqslant c$ as $y \to \infty$. Then

$$u(t) = \lim_{Y\to\infty} \frac{1}{2\pi i}\int_{c-iY}^{c+iY} \bar{u}(p) e^{pt}\,dp = u(t;b) + \sum_m u(t;\alpha_m) + \sum_m u(t;\beta_m),$$

where

$$u(t;b) = \lim_{Y\to\infty} \frac{1}{2\pi i}\int_{b-iY}^{b+iY} \bar{u}(p) e^{pt}\,dp,$$

$$u(t;\alpha_m) = \frac{1}{2\pi \mathrm{i}} \int^{(\alpha_m{}^+)} \bar{u}(p) \mathrm{e}^{pt} \, \mathrm{d}p,$$

$$u(t;\beta_m) = \frac{1}{2\pi \mathrm{i}} \int_{C_m} \bar{u}(p) \mathrm{e}^{pt} \, \mathrm{d}p,$$

and $C_m$ is a loop from $\mathscr{R}p = b$ round the cut from $\beta_m$ and back to $\mathscr{R}p = b$.

For

$$\int_{b+\mathrm{i}Y}^{c+\mathrm{i}Y} \bar{u}(p) \mathrm{e}^{pt} \, \mathrm{d}p \qquad \text{and} \qquad \int_{b-\mathrm{i}Y}^{c-\mathrm{i}Y} \bar{u}(p) \mathrm{e}^{pt} \, \mathrm{d}p$$

both tend to zero as $Y \to \infty$.

Note that the integrand in $u(t;b)$ is discontinuous (in general) on crossing the cuts.

**P3.4.4** Let $\bar{u}(p)$ satisfy the conditions of P3.4.3. Then as $t \to \infty$

(i) if

$$\int_{Y_0}^{\infty} [|\bar{u}(b+\mathrm{i}y)| + |\bar{u}(b-\mathrm{i}y)|] \, \mathrm{d}y$$

converges, $u(t;b) = \mathrm{o}(\mathrm{e}^{bt})$;

(ii) if

$$\int_{Y_0}^{\infty} [|\bar{u}'(b+\mathrm{i}y)| + |\bar{u}'(b-\mathrm{i}y)|] \, \mathrm{d}y$$

converges, $u(t;b) = \mathrm{o}(\mathrm{e}^{bt})$.

(i) $$|u(t;b)\mathrm{e}^{-bt}| = \frac{1}{2\pi} \left| \int_{-\infty}^{\infty} \bar{u}(b+\mathrm{i}y) \mathrm{e}^{\mathrm{i}yt} \, \mathrm{d}y \right| \leqslant \frac{1}{2\pi} \int_{-\infty}^{\infty} |\bar{u}(b+\mathrm{i}y)| \, \mathrm{d}y.$$

(ii) If $\bar{u}(b+\mathrm{i}y)$ is continuous for $|y| > Y_0$, integration by parts gives

$$\begin{aligned} u(t;b)\mathrm{e}^{-bt} = {} & \int_{-Y_0}^{Y_0} \bar{u}(b+\mathrm{i}y) \mathrm{e}^{\mathrm{i}yt} \, \mathrm{d}y - \int_{Y_0}^{\infty} \bar{u}'(b+\mathrm{i}y) \frac{\mathrm{e}^{\mathrm{i}yt}}{t} \, \mathrm{d}y \\ & - \int_{-\infty}^{-Y_0} \bar{u}'(b+\mathrm{i}y) \frac{\mathrm{e}^{\mathrm{i}yt}}{t} \, \mathrm{d}y - \frac{\bar{u}(b+\mathrm{i}Y_0)}{t} \exp(\mathrm{i}Y_0 t) \\ & + \frac{\bar{u}(b-\mathrm{i}Y_0)}{t} \exp(-\mathrm{i}Y_0 t). \end{aligned}$$

As $t \to \infty$, the first integral tends to zero by P3.2.3 and the others from the hypothesis (ii).

Note that if $\bar{u}(p) \sim Kp^{-\nu}$ throughout $|\arg p - \frac{1}{2}\pi| < \delta$ and $|\arg p + \frac{1}{2}\pi| < \delta$, then conditions (ii) hold if $\mathscr{R}\nu > 0$.

**P3.4.5** If

$$\bar{u}(p) = \sum_{n=0}^{\infty} b_n (p-\beta)^{n+\nu} \qquad \text{for } |p - \beta| < r$$

then

$$\mathrm{e}^{-\beta t} u(t;\beta) - \sum_{n=0}^{N-1} \frac{b_n}{\Gamma(-n-\nu)} t^{-n-\nu-1} = \mathrm{O}(t^{-N-\nu-1}) \quad \text{as } t \to \infty.$$

Without loss of generality, we can suppose that $\mathscr{R}\beta - \frac{1}{2}r > b$. Take $C_\beta$ to be made up of $C_2 + C_3 + C_4$, where $C_2$ and $C_4$ join $\mathscr{R}p = b$ to $\beta + \frac{1}{2}r\mathrm{e}^{\pm i\pi}$ and lie wholly in $b \leqslant \mathscr{R}p \leqslant \mathscr{R}\beta - \frac{1}{2}r$, and $C_3$ is a loop from $\beta + \frac{1}{2}r\mathrm{e}^{-i\pi}$ to $\beta + \frac{1}{2}r\mathrm{e}^{i\pi}$ lying in $|p - \beta| < r$ (Figure 3.5). From equation (3.3.1)

$$\frac{1}{2\pi i}\int_{-\infty}^{(\beta+)} (p-\beta)^{n+\nu} \mathrm{e}^{pt}\,\mathrm{d}p = \frac{t^{-n-\nu-1}}{\Gamma(-n-\nu)} \mathrm{e}^{\beta t}.$$

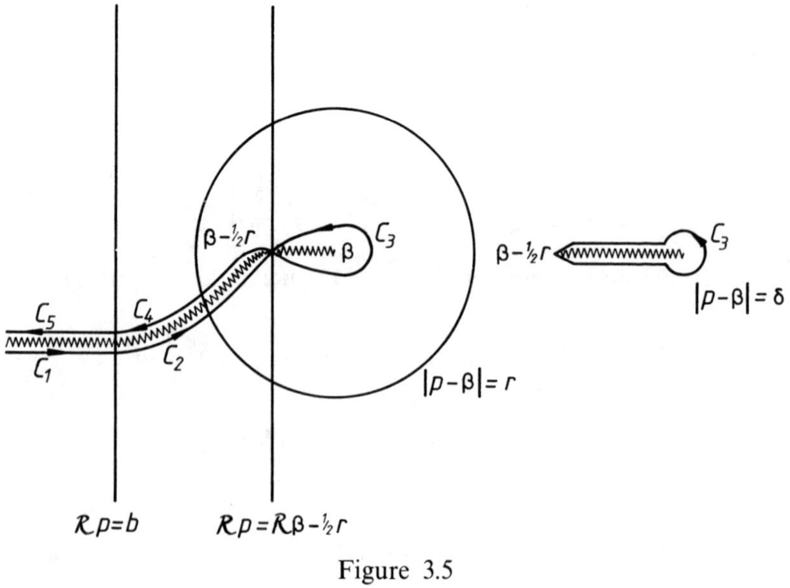


Figure 3.5

Then if $C_1$ and $C_5$ extend $C_\beta$ to $-\infty$,

$$2\pi i\left(u(t;\beta) - \sum_{n=0}^{N-1} \frac{b_n}{\Gamma(-n-v)} t^{-n-v-1} e^{\beta t}\right)$$

$$= \int_{C_\beta} \bar{u}_N(p) e^{pt} dp - \int_{C_1+C_5} \sum_{n=0}^{N-1} b_n (p-\beta)^{n+v} e^{pt} dp,$$

where

$$\bar{u}_N(p) = \bar{u}(p) - \sum_{n=0}^{N-1} b_n (p-\beta)^{n+v},$$

The method of P1.2.13 shows that as $t \to \infty$ the second integral is $O(e^{(b+\delta)t})$ for any $\delta > 0$. In the first integral $\bar{u}_N(p)$ is bounded on $C_2$ and $C_4$, so that

$$\int_{C_2+C_4} \bar{u}_N(p) e^{pt} dp = O(\exp[(\mathscr{R}\beta - \tfrac{1}{2}r)t]).$$

Now

$$f_N(p) = (p-\beta)^{-N-v} \bar{u}_N(p)$$

is regular in $|p-\beta| < r$ and so $|f_N(p)| < K$ on $C_3$. We can suppose that $\mathscr{R}(N+v) > -1$, because the result is stronger the greater $N$ is, and then we can take $C_3$ as $p = \beta + s e^{\pm i\pi}$ for $\delta \leqslant s \leqslant \frac{1}{2}r$ together with the circle $p = \beta + \delta e^{i\theta}$, and then make $\delta \to 0$. Hence

$$\int_{C_3} \bar{u}_N(p) e^{pt} dp = -2i \sin[(N+v)\pi] \int_0^{r/2} f(\beta - s) s^{N+v} e^{(\beta-s)t} ds$$

and so

$$\left|\int_{C_3}\right| < 2K |\sin v\pi| \exp(\mathscr{R}\beta t) \int_0^{r/2} s^{N+\mathscr{R}v} e^{-st} ds$$

$$< L \exp(\mathscr{R}\beta t) t^{-(N+\mathscr{R}v+1)}.$$

As with the case $t \to 0$, we can extend the argument to include logarithmic terms in the expansion of $\bar{u}(p)$ near the branch point. The formal result is obtained by differentiating that of P3.4.5 with respect to $v$.

**P3.4.6** If

$$\bar{u}(p) = \sum_{n=0}^{\infty} b_n (p-\beta)^{n+v} \log(p-\beta) \qquad \text{for } |p-\beta| < r$$

then as $t \to \infty$

$$e^{-\beta t}u(t;\beta) - \sum_{n=0}^{N-1} \frac{b_n}{\Gamma(-n-\nu)}[\psi(-n-\nu) - \log t]t^{-n-\nu-1}$$

$$= O(t^{-N-\nu-1}\log t).$$

The proof is analogous to that of P3.4.5.

It should be noted that although the formal asymptotic expansion of $u(t)$ as $t \to \infty$ is derived exclusively from those singularities with the greatest real part, the error inherent in cutting off an asymptotic series after a finite number of terms means that other singularities may be numerically significant when the value of $u(t)$ is required for some finite value of $t$. Because a pole gives only a finite number of terms, it is always worth considering an expansion as far as the branch points with greatest real part.

**E3.4.2** $\bar{u}(p) = p^{-1/2}(p-1)^{-1}$ has a simple pole at $p = 1$ and a branch point at $p = 0$. Its residue at the pole is 1, and near $p = 0$

$$\bar{u}(p) = -\sum_{n=0}^{\infty} p^{n-1/2} \qquad \text{for } |p| < 1.$$

The asymptotic expansion of $u(t)$ is therefore

$$u(t) \sim e^t - \sum_{n=0}^{\infty} \frac{1}{\Gamma(\frac{1}{2}-n)} t^{-n-1/2}$$

$$\sim e^t - (\pi t)^{-1/2} \sum_{n=0}^{\infty} (\tfrac{1}{2})_n(-t)^{-n}.$$

In fact

$$u(t) = (\pi t)^{-1/2} * e^t = e^t \operatorname{erf}(t^{1/2}).$$

E3.3.6 provides another example of this method, but in this case the expansion for large $t$ is a convergent series for $u_k(x, t)$. A necessary condition for this situation is that the expansion round the branch point $\beta$ of $\bar{u}(p)$ shall converge for all $p$, so that $p = \beta$ must be the only singularity of $\bar{u}(p)$.

**E3.4.3** Let $u(t) = J_\nu(t)$. Then, as shown in E2.4.3, if $\mathscr{R}\nu > -1$

$$(p^2+1)\frac{d^2\bar{u}}{dp^2} + 3p\frac{d\bar{u}}{dp} + (1-\nu^2)\bar{u} = 0,$$

from which

$$\bar{u}(p) = (p^2 + 1)^{-1/2}[(p^2 + 1)^{1/2} + p]^{-\nu}.$$

The only singularities of $\bar{u}(p)$ are branch points at $p = \pm \mathrm{i}$. If we put $p = \mathrm{i}(1 - 2q)$, the expansion of $\bar{u}(p)$ about $p = \mathrm{i}$ starts

$$\bar{u}(p) = \tfrac{1}{2}\exp(-\tfrac{1}{2}\nu\pi\mathrm{i})q^{-1/2}[1 + 2\nu\mathrm{i}q^{1/2} + (\tfrac{1}{2} - 2\nu^2)q + \mathrm{O}(q^{3/2})].$$

In terms of $q$, the differential equation becomes

$$q(q-1)\frac{\mathrm{d}^2\bar{u}}{\mathrm{d}q^2} - \tfrac{3}{2}(1 - 2q)\frac{\mathrm{d}\bar{u}}{\mathrm{d}q} + (1 - \nu^2)\bar{u} = 0.$$

This is a hypergeometric equation, and the solution with the given form near $q = 0$ is (see D1.5.2)

$$\bar{u}(p) = \tfrac{1}{2}\exp(-\tfrac{1}{2}\nu\pi\mathrm{i})[q^{-1/2}F(\tfrac{1}{2} + \nu, \tfrac{1}{2} - \nu; \tfrac{1}{2}; q) + 2\nu\mathrm{i}F(1 + \nu, 1 - \nu, \tfrac{3}{2}; q)].$$

The second term is regular at $q = 0$ and so does not contribute to the loop integral round $p = \mathrm{i}$. The first term gives the asymptotic expansion of this loop integral as

$$u(t;\mathrm{i}) \sim \tfrac{1}{2}\exp(-\tfrac{1}{2}\nu\pi\mathrm{i})\sum_{n=0}^{\infty}\frac{(\tfrac{1}{2} - \nu)_n(\tfrac{1}{2} - \nu)_n}{(\tfrac{1}{2})_n n!}\frac{\exp[(n - \tfrac{1}{2})\pi\mathrm{i}/2]t^{-n-1/2}}{2^{n-1/2}\Gamma(-n + \tfrac{1}{2})}\mathrm{e}^{\mathrm{i}t}$$

$$\sim \frac{\exp[-(\tfrac{1}{2}\nu + \tfrac{1}{4})\pi\mathrm{i}]}{(2\pi t)^{1/2}}\sum_{n=0}^{\infty}\frac{(\tfrac{1}{2} + \nu)_n(\tfrac{1}{2} - \nu)_n}{n!(2\mathrm{i}t)^n}\mathrm{e}^{\mathrm{i}t}.$$

Similarly, the loop round $p = -\mathrm{i}$ gives

$$u(t; -\mathrm{i}) \sim \frac{\exp[(\tfrac{1}{2}\nu + \tfrac{1}{4})\pi\mathrm{i}]}{(2\pi t)^{1/2}}\sum_{n=0}^{\infty}\frac{(\tfrac{1}{2} + \nu)_n(\tfrac{1}{2} - \nu)_n}{n!(-2\mathrm{i}t)^n}\mathrm{e}^{-\mathrm{i}t},$$

and therefore

$$\mathrm{J}_\nu(t) \sim \left(\frac{2}{\pi\tau}\right)^{1/2}\left(\cos(t - \tfrac{1}{2}\nu\pi - \tfrac{1}{4}\pi)\sum_{m=0}^{\infty}(-1)^m\frac{(\tfrac{1}{2} + \nu)_{2m}(\tfrac{1}{2} - \nu)_{2m}}{(2m)!(2t)^{2m}} + \sin(t - \tfrac{1}{2}\nu\pi - \tfrac{1}{4}\pi)\sum_{m=0}^{\infty}\frac{(-1)^m(\tfrac{1}{2} + \nu)_{2m+1}(\tfrac{1}{2} - \nu)_{2m+1}}{(2m+1)!(2t)^{2m+1}}\right)$$

as $t \to \infty$.

The method described above fails if $\bar{u}(p)$ has an essential singularity, or if it is regular for all $p$. In such cases it may be possible to use the saddle point method (method of steepest descent), described by Watson (1944, Section 8.3) and Olver (1974, Chapter 4, Section 7).

**E3.4.4** If $u(t) = t^{\nu/2} \mathrm{I}_\nu(2t^{1/2})$, where $\mathscr{R}\nu > -1$, then (E1.3.1) $\bar{u}(p) = p^{-\nu-1}\exp(p^{-1})$, which has an essential singularity at $p = 0$. From Jordan's lemma we get

$$t^{\nu/2}\mathrm{I}_\nu(2t^{1/2}) = \frac{1}{2\pi\mathrm{i}}\int_{-\infty}^{(0+)} p^{-\nu-1}\exp(pt + p^{-1})\mathrm{d}p.$$

In order to put this integral into a form suitable for the saddle point method, write $p = t^{1/2}z$. Then

$$\mathrm{I}_\nu(2t^{1/2}) = \frac{1}{2\pi\mathrm{i}}\int_{-\infty}^{(0+)} z^{-\nu-1}\exp[t^{1/2}f(z)]\mathrm{d}z,$$

where

$$f(z) = z + z^{-1}.$$

The saddle points are given by $f'(z) = 0$, so that $z = \pm 1$. On the surface of height $\mathscr{R}f(z)$, the paths of steepest descent through the saddle points are given by the real axis and the circle $|z| = 1$, the directions of descent being shown by the arrows in Figure 3.6. The path of integration is therefore taken along the lower edge of the cut from $\infty e^{-\pi i}$ to $e^{-\pi i}$, round the unit circle to $e^{\pi i}$, and then to $\infty e^{\pi i}$ along the upper edge of the cut. Since $f(1) = 2$ and $f(-1) = -2$, the dominant saddle point is at $z = 1$. With $\tau = 2t^{1/2}$, we then have

$$\mathrm{I}_\nu(\tau) = \frac{1}{2\pi}\int_{-\pi}^{\pi} e^{-\nu i\theta}\exp(\tau\cos\theta)\mathrm{d}\theta$$
$$-\frac{\sin\nu\pi}{\pi}\int_1^\infty r^{-\nu-1}\exp[-\tfrac{1}{2}\tau(r + r^{-1})]\mathrm{d}r,$$

where $z = e^{i\theta}$ on the unit circle and $z = re^{\pm i\pi}$ on the edges of the cut.

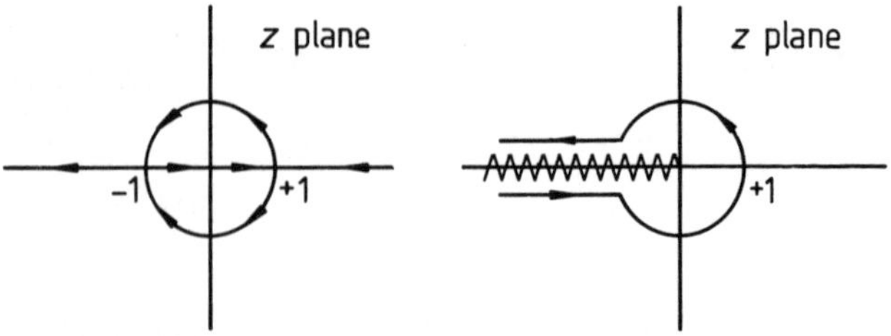


Figure 3.6

The second integral is $O(e^{-\tau})$ from P1.2.13. In the first integral put $s = \sin\frac{1}{2}\theta$ to obtain

$$I_\nu(\tau) = \frac{1}{\pi}\int_{-1}^{1} [(1-s^2)^{1/2} + is]^{-2\nu}$$
$$\times \exp[\tau(1-2s^2)](1-s^2)^{-1/2}ds + O(e^{-\tau})$$
$$= \frac{1}{\pi}\int_0^1 [g(s) + g(-s)]\exp[\tau(1-2s^2)]ds + O(e^{-\tau}),$$

where

$$g(s) = (1-s^2)^{-1/2}[(1-s^2)^{1/2} + is]^{-2\nu}.$$

The asymptotic expansion of this integral as $\tau \to \infty$ can now be obtained from P1.2.13 by taking $2s^2$ as the variable of integration.

The coefficients in the power series

$$g(s) = \sum_{n=0}^{\infty} a_n s^n$$

are most easily obtained from the differential equation

$$(1-s^2)g''(s) - 3sg'(s) + (4\nu^2 - 1)g(s) = 0.$$

The solution such that $g(0) = 1$ and $g'(0) = -2\nu i$ is

$$g(s) = F(\tfrac{1}{2}-\nu, \tfrac{1}{2}+\nu; \tfrac{1}{2}; s^2) - 2\nu i s F(1-\nu, 1+\nu; \tfrac{3}{2}; s^2).$$

We therefore obtain

$$I_\nu(\tau) \sim \frac{e^\tau}{(2\pi\tau)^{1/2}} \sum_{n=0}^{\infty} \frac{(\frac{1}{2}-\nu)_n(\frac{1}{2}+\nu)_n}{n!\,(2\tau)^n} \qquad \text{as } \tau \to \infty.$$

**E3.4.5** From E2.4.4, the Laplace transform of $u(t) = \mathrm{Ai}(t)$ is

$$\bar{u}(p) = \exp(-\tfrac{1}{3}p^3)\left[\int_0^p \exp(\tfrac{1}{3}z^3)\left(\frac{z}{3^{2/3}\Gamma(\frac{2}{3})} - \frac{1}{3^{1/3}\Gamma(\frac{1}{3})}\right)dz + \tfrac{1}{3}\right].$$

This function is regular for all $p$, and can be expressed as

$$\bar{u}(p) = \exp(-\tfrac{1}{3}p^3) - \exp(-\tfrac{1}{3}p^3)$$
$$\times \int_{-p}^{\infty} \exp(-\tfrac{1}{3}\zeta^3)\left(\frac{\zeta}{3^{2/3}\Gamma(\frac{2}{3})} + \frac{1}{3^{1/3}\Gamma(\frac{1}{3})}\right)d\zeta,$$

in which the second term is $O(p^{-1})$ as $p \to \infty$ in the left half plane. From Jordan's lemma, this term makes no contribution to the inverse transform when $t > 0$ and

$$\mathrm{Ai}(t) = \frac{1}{2\pi\mathrm{i}} \int_{\infty \exp(-2\pi\mathrm{i}/3)}^{\infty \exp(2\pi\mathrm{i}/3)} \exp(pt - \tfrac{1}{3}p^3)\mathrm{d}p.$$

For the saddle point method, put $p = t^{1/2}z$ so that

$$\mathrm{Ai}(t) = \frac{t^{1/2}}{2\pi\mathrm{i}} \int_{\infty \exp(-2\pi\mathrm{i}/3)}^{\infty \exp(2\pi\mathrm{i}/3)} \exp[t^{3/2}f(z)]\mathrm{d}z,$$

where $f(z) = z - \frac{1}{3}z^3$. The saddle points are at $z = \pm 1$ and if $z = x + \mathrm{i}y$ the paths of steepest descent through them are given by $y = 0$ or $x^2 - \frac{1}{3}y^2 = 1$. The directions of descent are shown by the arrows in Figure 3.7.

The required path of integration is the left branch of the hyperbola. On it, put $f(z) = f(-1) - \xi^2 = -\frac{2}{3} - \xi^2$. Then near the saddle point $z = -1 + \mathrm{i}\xi + \mathrm{O}(\xi^2)$, and the first approximation to $\mathrm{Ai}(t)$ as $t \to \infty$ is given by

$$\mathrm{Ai}(t) \sim \frac{t^{1/2}}{2\pi\mathrm{i}} \int_{-\infty}^{\infty} \exp[t^{3/2}(-\tfrac{2}{3} - \xi^2)]\mathrm{i}\mathrm{d}\xi$$

$$\sim \frac{1}{2\sqrt{\pi}} t^{-1/4} \exp(-\tfrac{2}{3}t^{3/2}).$$

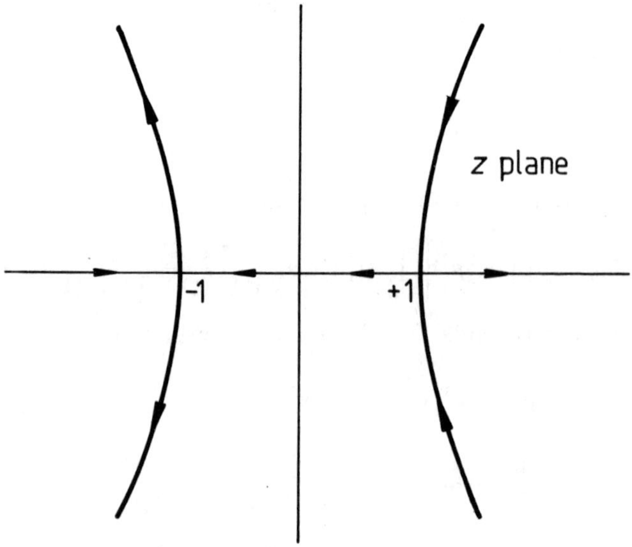


Figure 3.7

In order to extend this into an asymptotic expansion, we need to express $z$ as a power series in $\xi$. Near $\xi = 0$, $z$ is a regular function of $\xi$, so that

$$z = \sum_{n=0}^{\infty} a_n \xi^n,$$

where

$$a_n = \frac{1}{2\pi i} \int^{(0+)} \frac{z}{\xi^{n+1}} d\xi = \frac{1}{2\pi i} \int^{(-1+)} \frac{z}{\xi^{n+1}} \frac{d\xi}{dz} dz.$$

Since

$$2\xi \frac{d\xi}{dz} = -f'(z) = z^2 - 1$$

and

$$\xi = -i(z+1)\left(\frac{2-z}{3}\right)^{1/2}$$

we obtain

$$a_n = \frac{1}{2\pi i} \int^{(0+)} (-\tfrac{1}{2}i^n)(1-\zeta)(2-\zeta)(1-\tfrac{1}{3}\zeta)^{-n/2-1} \zeta^{-n-1} d\zeta,$$

where $\zeta = z + 1$. Consequently, $a_n$ is the coefficient of $\zeta^n$ in

$$-\tfrac{1}{2}i^n(1-\zeta)(2-\zeta)(1-\tfrac{1}{3}\zeta)^{-n/2-1},$$

and hence

$$a_n = \frac{i^n}{3^{n-1}} \frac{(\frac{1}{2}n)_{n-1}}{n!}.$$

Thus

$$\begin{aligned} \mathrm{Ai}(t) &= \frac{t^{1/2}}{2\pi i} \int_{-\infty}^{\infty} \exp[t^{3/2}(-\tfrac{2}{3} - \xi^2)] \frac{dz}{d\xi} d\xi \\ &\sim \frac{t^{1/2}}{2\pi i} \sum_{m=0}^{\infty} (2m+1) a_{2m+1} \int_{-\infty}^{\infty} \exp[t^{3/2}(-\tfrac{2}{3} - \xi^2)] \xi^{2m} d\xi. \end{aligned}$$

After some reduction, this gives

$$\mathrm{Ai}(t) \sim \frac{\exp(-\frac{2}{3}t^{3/2})}{2\sqrt{\pi} t^{1/4}} \sum_{m=0}^{\infty} \frac{(\frac{1}{6})_m (\frac{5}{6})_m}{m!} \left(-\frac{3}{4t^{3/2}}\right)^m.$$

If an inverse transform cannot be evaluated effectively by means of convergent or asymptotic expansions, numerical integration will be needed. Loop integrals are often useful because the factor $e^{pt}$ in the integrand vanishes as $\mathscr{R}p \to -\infty$. In choosing a contour for integration, care should be taken to keep clear of singularities where the transform is large, and to avoid rapid oscillations of the integrand. The ideas used in the asymptotic analyses are helpful even when $t$ has a moderate value, of order 1. A method of numerical inversion by means of loop integrals has been developed by Talbot (1979).

## 3.5 Additional Examples

**E3.5.1** Use the residue method to find the inverse transforms of

(i) $\dfrac{1}{p(p+1)(p+2)}$,

(ii) $\dfrac{p+1}{p^2(p-1)}$,

(iii) $\dfrac{1}{p^3(p+1)^2}$,

(iv) $\dfrac{p+1}{p(p^2+1)^2}$.

**E3.5.2** Show that if $n = 0, 1, 2, \ldots$

$$\mathscr{L}^{-1}\left\{\frac{1}{p^{n+1}(p+1)^{n+1}}\right\} = \frac{(-1)^n}{n!}\sum_{m=0}^{n}\frac{(2n-m)!}{(n-m)!\,m!}t^m[(-1)^m - e^{-t}].$$

**E3.5.3** Use the residue method to obtain series for the inverse transforms of

(i) $\dfrac{\Gamma(p+\alpha)}{\Gamma(p+\beta)}$ $\quad(\mathscr{R}\beta > \mathscr{R}\alpha)$,

(ii) $\psi(p+\alpha) - \psi(p+\beta)$,

(iii) $\operatorname{sech}(\sqrt{p})$,

(iv) $\operatorname{cosech}^2(\sqrt{p})$.

**E3.5.4** Sketch graphs of the inverse transforms of

(i) $p^{-2}\operatorname{sech}^2 p$,

(ii) $\dfrac{1}{(p^2+1)(1-e^{-\pi p})}$.

**E3.5.5** (i) Show that

$$\mathscr{L}^{-1}\{\operatorname{sech}(\surd p)\}$$
$$= \pi^{-1/2} t^{-3/2} \sum_{n=0}^{\infty} (-1)^n (2n+1) \exp[-(n+\tfrac{1}{2})^2/t].$$

(ii) Find an analogous expression for $\mathscr{L}^{-1}\{\operatorname{cosech}^2(\surd p)\}$.

**E3.5.6** Show that

$$\bar{u}(p, \alpha) = \frac{\cosh[(1-\alpha)\surd p]}{\surd(p)\sinh(\surd p)}$$

is a Laplace transform if $|\arg \alpha|$ and $|\arg(2-\alpha)|$ are both less than $\frac{1}{4}\pi$, and that its inverse transform is

$$u(t, \alpha) = (\pi t)^{-1/2} \sum_{n=-\infty}^{\infty} \exp[-(n+\tfrac{1}{2}\alpha)^2/t].$$

Prove that if $\alpha$ is real, and $0 \leqslant \alpha \leqslant 2$, then

$$u(t, \alpha) = 1 + 2 \sum_{n=1}^{\infty} \cos(n\pi\alpha) \exp(-n^2\pi^2 t).$$

[**D3.5.1**

$$\vartheta_3(v|\tau) = 1 + 2 \sum_{n=1}^{\infty} \cos(2n\pi v) \exp(\mathrm{i}n^2\pi\tau),$$

where $\mathscr{I}\tau > 0$, is one of the *theta functions*. E3.5.6 shows that

$$u(t, \alpha) = \vartheta_3(\tfrac{1}{2}\alpha|\mathrm{i}\pi t) = (\pi t)^{-1/2} \exp\left(-\frac{\alpha^2}{4t}\right) \vartheta_3\left(\frac{\mathrm{i}\alpha}{2\pi t}\middle|\frac{\mathrm{i}}{\pi t}\right),$$

and analytic continuation shows that this transformation of the theta function is valid for all complex values of $\alpha$ and $t$ with $\mathscr{R}t > 0$. For application see E4.3.2.]

**E3.5.7** Show that $\bar{u}(p) = \exp(-p^\nu)$ is a Laplace transform if $\nu$ is real with $0 < \nu < 1$. Express the inverse transform $u(t)$ as a real integral, and in particular show that

$$\mathscr{L}^{-1}\{\exp(-\surd p)\} = \frac{1}{\pi} \int_0^\infty \sin(\surd r) e^{-rt} \, dr.$$

**E3.5.8** Given that

$$\bar{u}(p) = \frac{\exp[-a\surd(p^2 - \mathrm{K}^2)]}{\surd(p^2 - \kappa^2)},$$

where $a > 0$, evaluate its inverse transform as follows.
(i) Show that

$$u(t) = \frac{1}{2\pi \mathrm{i}} \int_C \frac{\exp[pt - a\surd(p^2 - \kappa^2)]}{\surd(p^2 - \kappa^2)} \mathrm{d}p \, \mathrm{H}(t - a),$$

where the contour $C$ encloses the branch points $p = \pm\kappa$.
(ii) Transform the integral by the substitution $p = \kappa \cosh \zeta$ into

$$u(t) = \frac{1}{2\pi \mathrm{i}} \int_{\alpha - \mathrm{i}\pi}^{\alpha + \mathrm{i}\pi} \exp(\kappa t \cosh \zeta - \kappa a \sinh \zeta) \mathrm{d}\zeta \, \mathrm{H}(t - a),$$

where $\alpha$ is arbitrary.
(iii) Choose $\alpha = \theta$, where $t = a \coth \theta$, and put $\zeta = \theta + i\eta$ to get

$$u(t) = \frac{1}{2\pi} \int_{-\pi}^{\pi} \exp(\kappa a \operatorname{cosech} \theta \cos \eta) \mathrm{d}\eta \, \mathrm{H}(t - a).$$

(iv) Evaluate the integral and obtain

$$u(t) = \mathrm{I}_0(\kappa\surd(t^2 - a^2)) \mathrm{H}(t - a).$$

**E3.5.9** If $\bar{u}(p) = (p^{3/2} - 1)^{-1}$, show that

$$u(t) = \tfrac{2}{3}\mathrm{e}^t - \frac{1}{\pi} \int_0^\infty \frac{r^{3/2}}{r^3 + 1} \mathrm{e}^{-rt} \mathrm{d}r.$$

Hence find the asymptotic expansion of $u(t)$ as $t \to \infty$. Obtain a convergent series for $u(t)$ by expanding $\bar{u}(p)$ as a descending series. [$u(t)$ may be expressed in terms of error functions with the aid of E1.5.26 (ii).]

**E3.5.10** If $\bar{u}(p) = (p^2 + 1)^{-1} \log p$, show that

$$u(t) = \tfrac{1}{2}\pi \cos t - \int_0^\infty (r^2 + 1)^{-1} \mathrm{e}^{-rt} \mathrm{d}r.$$

Find an asymptotic series for $u(t)$ as $t \to \infty$, and a convergent series for $u(t)$ involving $\log t$ and ascending powers of $t$. [$u(t)$ may be evaluated in terms of sine and cosine integrals by means of E1.3.8.]

**E3.5.11** If $\bar{u}(p) = p^{-1}\operatorname{cosech}(\sqrt{p})$, show that

$$u(t) = 2\sum_{n=0}^{\infty} \operatorname{erfc}((n+\tfrac{1}{2})/\sqrt{t}),$$

and obtain the asymptotic expansion of $u(t)$ as $t \to \infty$. [Use

$$\frac{1}{\sinh x} = \frac{2}{1-e^{-x}} - \frac{2}{1-e^{-2x}}$$

and D1.5.6.]

**E3.5.12** Show that

$$\mathscr{L}^{-1}\{p^{\mu}K_{\nu}(2(\alpha p)^{1/2})\} = \tfrac{1}{2}t^{\mu-1}e^{-\eta}\eta^{\nu/2}U(\tfrac{1}{2}\nu-\mu; \nu+1; \eta),$$

where $\mathscr{R}\alpha > 0$ and $\eta = \alpha/t$. In particular, obtain

(i) $\mathscr{L}^{-1}\{p^{-\nu/2}K_{\nu}(2(\alpha p)^{1/2})\} = \tfrac{1}{2}\alpha^{\nu/2}t^{-\nu-1}e^{-\alpha/t}$,

(ii) $\mathscr{L}^{-1}\{p^{-1/2}K_{\nu}(2(\alpha p)^{1/2})\} = \tfrac{1}{2}\pi^{-1/2}t^{-3/2}e^{-\eta/2}K_{\nu/2}(\tfrac{1}{2}\eta)$.

**E3.5.13** Prove that if $\mathscr{R}\nu > -1$ and $0 < x \leqslant y$,

$$\mathscr{L}^{-1}\{I_{\nu}(x\sqrt{p})K_{\nu}(y\sqrt{p})\} = \tfrac{1}{2}\int_0^{\infty} J_{\nu}(x\sqrt{r})J_{\nu}(y\sqrt{r})e^{-rt}\,dr,$$

and evaluate the result by means of E1.5.37.

**E3.5.14** (i) Find a series expansion for

$$u(t) = \mathscr{L}^{-1}\{p^{-1}\exp(p^{-1/2})\},$$

and apply the saddle point method to show that as $t \to \infty$

$$u(t) \sim (3\pi)^{-1/2}(\tfrac{1}{4}t)^{-1/6}\exp[3(\tfrac{1}{4}t)^{1/3}].$$

(ii) Obtain the corresponding results for $\mathscr{L}^{-1}\{p^{-1}\exp(-p^{-1/2})\}$.

**E3.5.15** Given that $\bar{u}(p) = \exp(-p^{\nu})$, where $0 < \nu < 1$, show that

$$u(t) = \frac{1}{\pi}\sum_{n=1}^{\infty}\frac{(-1)^{n-1}\sin(n\pi\nu)\Gamma(n\nu+1)}{n!\,t^{n\nu+1}}.$$

Use the saddle point method to show that as $t \to 0+$

$$u(t) \sim [2\pi\nu(1-\nu)]^{-1/2}\left(\frac{\nu}{t}\right)^{(2-\nu)/(2-2\nu)} \times \exp\left[-(1-\nu)\left(\frac{\nu}{t}\right)^{\nu/(1-\nu)}\right].$$

**E3.5.16** Show that

$$u(t) = \sum_{m,n=1}^{\infty} \frac{\sin(mt/n)}{m^2 n^2}$$

has a Laplace transform for $\mathscr{R}p > 0$. Prove that as $\delta \to 0+$

$$\bar{u}\left(\mathrm{i}\frac{k}{l} + \delta\right) = -\frac{\pi^4 \mathrm{i}}{180k^2 l^2 \delta} + \mathrm{O}(1),$$

where $k$ and $l$ are positive co-prime integers, and that

$$\bar{u}(\delta) = -\zeta(3)\log\delta + \mathrm{O}(1).$$

Deduce that $\mathscr{R}p = 0$ is a natural boundary for $\bar{u}(p)$.

**E3.5.17** (i) Prove that if $f(p)\mathrm{e}^{xp}$ is a Laplace transform for all $x \geqslant 0$, then $f(p) \equiv 0$.

[If $f(p)\mathrm{e}^{Xp} = \bar{u}(p)$, then $f(p) = \mathscr{L}\{u(t-X)\,\mathrm{H}(t-X)\}$.]

(ii) Prove that if $g(p)\mathrm{I}_\nu(xp)$, where $\nu$ is real, is a Laplace transform for all $x \geqslant a > 0$, then $g(p) \equiv 0$.

[Use the fact that

$$\frac{\mathrm{I}_\nu(ap)}{\mathrm{I}_\nu(Xp)}\mathrm{e}^{(X-a)p}$$

is bounded in $\mathscr{R}p \geqslant A > 0$.]

**E3.5.18** Show that the Laplace transform of $u(t) = \cos(t^2)$ satisfies the equation

$$\bar{u}(\mathrm{i}p) = \tfrac{1}{2}\sqrt{\pi}\exp\left[\mathrm{i}\left(\tfrac{1}{4}\pi - \tfrac{1}{4}p^2\right)\right] - \mathrm{i}\bar{u}(p).$$

Deduce that if $x \leqslant 0$, $\bar{u}(x + \mathrm{i}y)$ does not tend to zero as $y \to \pm\infty$.

**E3.5.19** Prove that if

$$\int_0^\infty (1+t)^{-1}|u(t)|\,\mathrm{d}t$$

converges, the Stieltjes transform (see E1.5.33 (iii))

$$\bar{\bar{u}}(p) = \int_0^\infty \frac{u(t)}{p+t}\,\mathrm{d}t$$

is regular in the $p$ plane cut along the negative real axis from $-\infty$ to 0.

If $u(\tau \pm 0)$ exist such that as $s \to 0+$

$$|u(\tau \pm s) - u(\tau \pm 0)| = O(s^a)$$

where $a > 0$, prove that

$$\lim_{\delta \to 0+} \frac{1}{2\pi i}[\bar{\bar{u}}(-\tau - i\delta) - \bar{\bar{u}}(-\tau + i\delta)] = \tfrac{1}{2}[u(\tau + 0) + u(\tau - 0)].$$

**E3.5.10** Use E3.5.19 and E1.5.33 (iii) to find the functions $u(t)$ such that $\bar{u}(t) = \lambda u(t)$, where $\lambda$ is a constant.

# 4
# Partial Differential Equations

## 4.1 First-Order Equations: Types of Second-Order Equations

In Chapter 2, the Laplace transformation was used to reduce ordinary differential equations to algebraic equations. In the same way, a partial differential equation in two variables $x$ and $t$ may be reduced to an ordinary differential equation in $x$ by means of a Laplace transformation with respect to $t$. For this the equation must be linear and the coefficients of the unknown function and its derivatives must be independent of $t$. Consider a simple first-order equation to illustrate the method.

**E4.1.1**

$$\frac{\partial u}{\partial t} + a\frac{\partial u}{\partial x} = bu + ct,$$

where $a, b$ and $c$ are constants, with $a > 0$. We wish to solve this equation in the region $x > 0, t > 0$, given that

$$u(x, 0) = \chi(x) \qquad \text{(initial condition)},$$

$$u(0, t) = g(t) \qquad \text{(boundary condition)}.$$

The Laplace transform

$$\bar{u}(x, p) = \int_0^\infty u(x, t)e^{-pt}\,dt$$

satisfies

$$p\bar{u} - \chi(x) + a\frac{\partial \bar{u}}{\partial x} = b\bar{u} + cp^{-2},$$

assuming that we can differentiate with respect to $x$ under the integral sign, and

$$\bar{u}(0, p) = \bar{g}(p).$$

After multiplying by the appropriate integrating factor, we can integrate to obtain

$$\bar{u}(x, p) = \frac{1}{a}\int_0^x \exp\left(\frac{p-b}{a}(y-x)\right)\chi(y)\,\mathrm{d}y + \frac{c}{p^2(p-b)}$$
$$+ A(p)\exp\left(-\frac{p-b}{a}x\right).$$

The 'constant of integration' is the function $A(p)$, given by the boundary condition as

$$A(p) = \bar{g}(p) - cp^{-2}(p-b)^{-1}.$$

If the functions $\chi(x)$ and $g(t)$ are suitably chosen, we can work out the transform $\bar{u}(x,p)$ explicitly and invert it to obtain the solution $u(x, t)$. However, it is instructive to consider the general case to see how the solution depends on the choice of these functions. We have

$$\bar{u}(x, p) = a^{-1}\int_0^x \chi(y)\exp\left(-(p-b)\frac{x-y}{a}\right)\mathrm{d}y$$
$$+ \bar{g}(p)\mathrm{e}^{-(p-b)x/a} + cp^{-2}(p-b)^{-1}[1 - \mathrm{e}^{-(p-b)x/a}].$$

The second and third terms are the transform of

$$g\left(t-\frac{x}{a}\right)\mathrm{e}^{bx/a}\mathrm{H}\left(t-\frac{x}{a}\right)$$
$$+\frac{c}{b^2}\left\{\mathrm{e}^{bt} - 1 - bt - \mathrm{e}^{bx/a}\left[\mathrm{e}^{b(t-x/a)} - 1 - b\left(t-\frac{x}{a}\right)\right]\mathrm{H}\left(t-\frac{x}{a}\right)\right\}.$$

The first term, on writing $y = x - as$, becomes

$$\int_0^{x/a} \chi(x-as)\mathrm{e}^{bs-ps}\,\mathrm{d}s,$$

and so is the Laplace transform of

$$\chi(x-at)\mathrm{e}^{bt}\mathrm{H}(x-at).$$

Thus the general solution is

$$u(x, t) = cb^{-2}(\mathrm{e}^{bt} - 1 - bt)$$
$$+\left\{g\left(t-\frac{x}{a}\right) - \frac{c}{b^2}\left[\mathrm{e}^{b(t-x/a)} - 1 - b\left(t-\frac{x}{a}\right)\right]\right\}$$
$$\times\,\mathrm{e}^{bx/a}\mathrm{H}\left(t-\frac{x}{a}\right) + \chi(x-at)\mathrm{e}^{bt}\mathrm{H}(x-at).$$

The solution in $x > at$ depends on $\chi(x)$ but not on $g(t)$; in $t > x/a$ it depends on $g(t)$ but not on $\chi(x)$. The values of these functions are carried along the lines $x - at = \text{constant}$, which are the *characteristics* of the differential equation. If $\chi(0) \neq g(0)$, or if the functions $\chi(x)$ and $g(t)$ are not continuous, the solution will be discontinuous along a characteristic, and the differential equation will not hold on this line.

**D4.1.1** The curve $C$, given parametrically as $(x(s), t(s))$ is a characteristic of the partial differential equation

$$P(x, t, u)\frac{\partial u}{\partial x} + Q(x, t, u)\frac{\partial u}{\partial t} = R(x, t, u) \tag{4.1.1}$$

if the equation relating the differentials along $C$, namely

$$\frac{\mathrm{d}x}{\mathrm{d}s}\frac{\partial u}{\partial x} + \frac{\mathrm{d}t}{\mathrm{d}s}\frac{\partial u}{\partial t} = \frac{\mathrm{d}u}{\mathrm{d}s},$$

is linearly dependent on equation (4.1.1).

The conditions for a characteristic are therefore

$$\frac{\mathrm{d}x}{P} = \frac{\mathrm{d}t}{Q} = \frac{\mathrm{d}u}{R}. \tag{4.1.2}$$

Equation (4.1.1) asserts geometrically that the surface $u = u(x, t)$ in $x, t, u$ space is parallel to the vector field with components $(P, Q, R)$ and so the lines of flow of this field, which are given by (4.1.2), lie in the surface.

For a linear equation $P$ and $Q$ are independent of $u$, so that the characteristics are independent of the particular solution $u(x, t)$. The second equality of (4.1.2) is then an equation for the variation of $u$ along each fixed characteristic.

**E4.1.1** (continued) If we apply this method to our example, equation (4.1.2) becomes

$$\frac{\mathrm{d}x}{a} = \frac{\mathrm{d}t}{1} = \frac{\mathrm{d}u}{bu + ct}.$$

Thus on $\mathrm{d}x/\mathrm{d}t = a$ we have $\mathrm{d}u/\mathrm{d}t = bu + ct$, so that on any line $x - at = A$ we have $u = Be^{bt} - ct/b - c/b^2$, where $A$ and $B$ are constants. The particular solution we want is found by choosing a functional relation between $A$ and $B$ so that

$$u(x, 0) = \chi(x), \qquad u(0, t) = g(t)$$

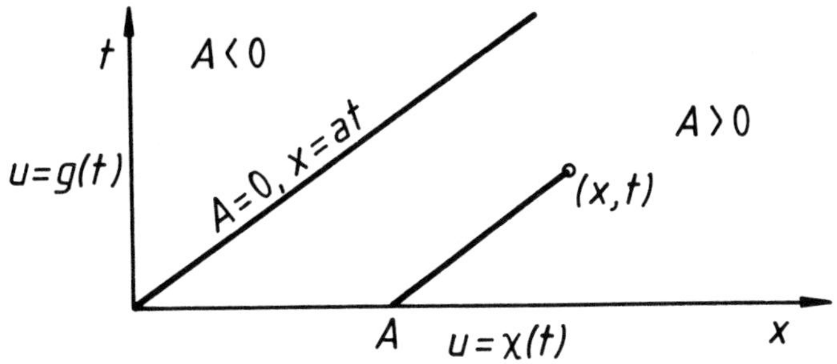


Figure 4.1

Note that $A>0$ on the positive $x$ axis, and $A<0$ on the positive $t$ axis (see Figure 4.1). If $A>0$, we have $x=A$ when $t=0$ so that $\chi(A)=B-c/b^2$. Thus for $A>0$

$$u=[\chi(A)+c/b^2]e^{bt}-ct/b-c/b^2,$$

that is

$$u(x,t)=[\chi(x-at)+c/b^2]e^{bt}-ct/b-c/b^2 \qquad \text{when } x>at.$$

If $A<0$, $t=-A/a$ when $x=0$ so that

$$g(-A/a)=Bc^{-Ab/a}+\frac{cA}{ba}-\frac{c}{b^2},$$

and

$$B=\left[g\left(-\frac{A}{a}\right)+\frac{c}{b}\left(-\frac{A}{a}\right)+\frac{c}{b^2}\right]e^{Ab/a}.$$

Hence when $t>x/a$,

$$u(x,t)=\left[g\left(t-\frac{x}{a}\right)+\frac{c}{b}\left(t-\frac{x}{a}\right)+\frac{c}{b^2}\right]e^{bx/a}-\frac{ct}{b}-\frac{c}{b^2}.$$

In this example we must have $a>0$, since if $a\leqslant 0$ the characteristics are vertical or have negative slope, and the functions $g(t)$, $\chi(x)$ cannot be chosen independently.

The Laplace transformation is more useful for second-order than for first-order partial differential equations. These second-order equations are classified according to their characteristics.

**D4.1.2** The curve $C$, given parametrically as $(x(s), t(s))$, is a characteristic of the partial differential equation

$$P\frac{\partial^2 u}{\partial x^2}+2Q\frac{\partial^2 u}{\partial x\,\partial t}+R\frac{\partial^2 u}{\partial t^2}=S, \tag{4.1.3}$$

where $P, Q, R$ and $S$ are functions of $x, t, u, \partial u/\partial x$ and $\partial u/\partial t$ if the equations

$$\frac{dx}{ds}\frac{\partial^2 u}{\partial x^2}+\frac{dt}{ds}\frac{\partial^2 u}{\partial x\partial t}=\frac{d}{ds}\left(\frac{\partial u}{\partial x}\right),$$

$$\frac{dx}{ds}\frac{\partial^2 u}{\partial x\partial t}+\frac{dt}{ds}\frac{\partial^2 u}{\partial t^2}=\frac{d}{ds}\left(\frac{\partial u}{\partial t}\right),$$

which hold on $C$, form a linearly dependent set with equation (4.1.3).

The characteristic directions are therefore given by

$$P\left(\frac{dt}{ds}\right)^2-2Q\frac{dx}{ds}\frac{dt}{ds}+R\left(\frac{dx}{ds}\right)^2=0,$$

so that

$$\frac{dx}{dt}=R^{-1}[Q\pm(Q^2-PR)^{1/2}]. \tag{4.1.4}$$

Corresponding to each characteristic direction there is a relation between

$$\frac{d}{ds}\left(\frac{\partial u}{\partial x}\right) \quad \text{and} \quad \frac{d}{ds}\left(\frac{\partial u}{\partial t}\right).$$

If the coefficients $P$, $Q$ and $R$ are functions of $x$ and $t$ only, as in a linear equation, the characteristics are independent of the solution $u(x, t)$, so that the two families of characteristics can be calculated in advance by integrating equation (4.1.4). When $P, Q$ and $R$ are real, the characteristic directions are real and distinct if $Q^2 > PR$, coincident if $Q^2 = PR$, and complex if $Q^2 < PR$.

**D4.1.3** The partial differential equation (4.1.3) with $P$, $Q$ and $R$ real is

$$\left\{\begin{matrix}\text{hyperbolic}\\ \text{parabolic}\\ \text{elliptic}\end{matrix}\right\} \text{in the region } D \text{ if } Q^2 \left\{\begin{matrix}>\\ =\\ <\end{matrix}\right\} PR \text{ throughout } D.$$

The typical examples of the three types of second-order equations are the following.

(1) Wave equation

$$\frac{\partial^2 u}{\partial t^2}=c^2\frac{\partial^2 u}{\partial x^2},$$

where $c$ is a positive constant. This is hyperbolic, with characteristics $x \pm ct = \text{constant}$.

(2) Diffusion or heat-conduction equation

$$\frac{\partial u}{\partial t} = \kappa \frac{\partial^2 u}{\partial x^2},$$

where $\kappa$ is a positive constant. This is parabolic with characteristics $t = \text{constant}$.

(3) Laplace's equation

$$\frac{\partial^2 u}{\partial x^2} + \frac{\partial^2 u}{\partial y^2} = 0.$$

This is elliptic, with characteristics $x \pm \mathrm{i}y = \text{constant}$.

The generalizations of these equations to three space dimensions are respectively

(1) $\dfrac{\partial^2 u}{\partial t^2} = c^2 \nabla^2 u,$

(2) $\dfrac{\partial u}{\partial t} = \kappa \nabla^2 u,$

(3) $\nabla^2 u = 0,$

where

$$\nabla^2 u = \frac{\partial^2 u}{\partial x^2} + \frac{\partial^2 u}{\partial y^2} + \frac{\partial^2 u}{\partial z^2}.$$

The Laplace transformation is often useful for linear second-order equations when these are hyperbolic or parabolic, but rarely for elliptic equations. This is because a well posed problem for an elliptic equation needs a condition all round a closed boundary, whereas for hyperbolic or parabolic equations initial conditions and side boundary conditions must be given. Typical boundary conditions are shown in Figure 4.2.

## 4.2 The Wave Equation

As noted above, the two families of characteristics of the wave equation (in which $c$ is a positive constant)

$$\frac{\partial^2 u}{\partial t^2} = c^2 \frac{\partial^2 u}{\partial x^2}$$

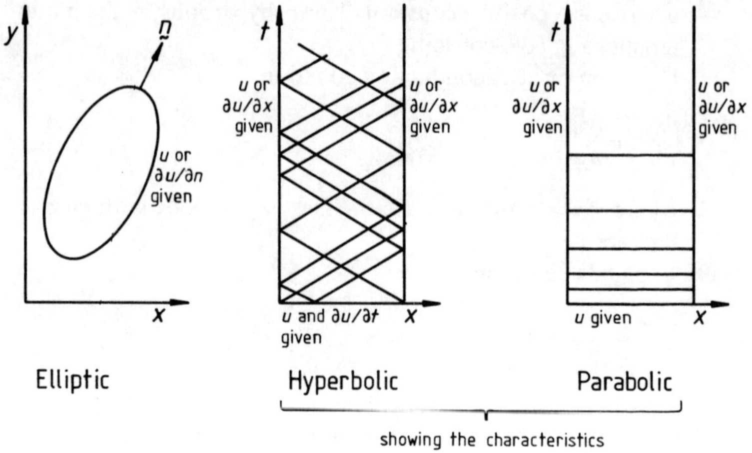


Figure 4.2

are $\xi = \text{constant}$, $\eta = \text{constant}$, where

$$\xi = x - ct, \qquad \eta = x + ct.$$

If $\xi, \eta$ are taken as independent variables, the wave equation becomes

$$\frac{\partial^2 u}{\partial \xi \partial \eta} = 0,$$

and therefore has the general solution

$$u = f(\xi) + g(\eta) = f(x - ct) + g(x + ct),$$

where $f$ and $g$ denote arbitrary functions. This is *d'Alembert's solution*; the first term represents a disturbance travelling with speed $c$ in the positive $x$ direction, the second term a disturbance travelling in the negative $x$ direction, also with speed $c$. For this reason, $c$ is called the *wave speed*.

The wave equation is the governing equation for small disturbances on stretched strings and acoustic vibrations in tubes. The three-dimensional form $\partial^2 u/\partial t^2 = c^2 \nabla^2 u$ governs general acoustic vibrations and also occurs in the theories of elastic and electromagnetic waves. Since it contains $\partial^2 u/\partial t^2$ we require two initial conditions, the values of $u$ and $\partial u/\partial t$ at $t = 0$. The condition at the boundaries of the appropriate region of space may be on either $u$ or its normal derivative.

**E4.2.1**

$$\frac{\partial^2 u}{\partial t^2} = c^2 \frac{\partial^2 u}{\partial x^2} \qquad \text{in } x > 0, t > 0$$

with

$$u = \chi(x), \qquad \partial u/\partial t = 0 \qquad \text{at } t = 0,$$

$$u = g(t) \qquad \text{at } x = 0.$$

The Laplace transform of the equation is

$$p^2\bar{u} - p\chi(x) = c^2 \frac{\partial^2 \bar{u}}{\partial x^2},$$

and we also have $u(0, p) = \bar{g}(p)$. Since this is a second-order equation for $\bar{u}(x, p)$ we need another condition. This comes from the principle (E3.5.17) that if $f(p)e^{xp}$ is a Laplace transform for all $x \geqslant 0$ (or all $x \geqslant a$), then $f(p) \equiv 0$.

The solution for $\bar{u}(x, p)$, obtained either by a Laplace transform with respect to $x$ or by variation of parameters, is

$$\bar{u}(x, p) = \bar{g}(p) \cosh\left(\frac{px}{c}\right) + \frac{c}{p}\bar{\psi}(p) \sinh\left(\frac{px}{c}\right) - \frac{1}{c} \sinh\left(\frac{px}{c}\right) * \chi(x),$$

where

$$\psi(t) = \frac{\partial u}{\partial x}(0, t).$$

This may be put in the form

$$\bar{u}(x, p) = \frac{1}{2c}\int_x^\infty \chi(y)e^{p(x-y)/c}\,dy + \frac{1}{2c}\int_0^x \chi(y)e^{p(y-x)/c}\,dy$$

$$+ \frac{1}{2}\left(\bar{g}(p) - \frac{c}{2p}\bar{\psi}(p)\right)e^{-px/c}$$

$$+ \frac{1}{2}\left[\bar{g}(p) + \frac{c}{2p}\bar{\psi}(p) - \frac{1}{c}\bar{\chi}\left(\frac{p}{c}\right)\right]e^{px/c},$$

where we assume that the function $\chi(t)$ has a Laplace transform. The first two terms can be expressed as Laplace transforms by means of the substitutions $y = x \pm cs$, and the third term is also a transform. Hence, by the principle mentioned, the last term must vanish identically, so that

$$\bar{u}(x,p) = \frac{1}{2}\int_0^\infty \chi(x+cs)\mathrm{e}^{-ps}\,\mathrm{d}s + \frac{1}{2}\int_0^{x/c} \chi(x-cs)\mathrm{e}^{-ps}\,\mathrm{d}s$$
$$+\left[\bar{g}(p) - \frac{1}{2c}\bar{\chi}\left(\frac{p}{c}\right)\right]\mathrm{e}^{-px/c}.$$

This is the Laplace transform of

$$u(x,t) = \tfrac{1}{2}\chi(x+ct) + \tfrac{1}{2}\chi(x-ct)\mathrm{H}(x-ct)$$
$$+\left[g\left(t-\frac{x}{c}\right) - \frac{1}{2}\chi(ct-x)\right]\mathrm{H}\left(t-\frac{x}{c}\right).$$

The answer is in the form of d'Alembert's solution, and shows how the values imposed at $t=0$ and at $x=0$ propagate along the characteristics, with reflexion at $x=0$. If there is more than one boundary, there will be repeated reflexion of the characteristics, as shown in the next example.

**E4.2.2**

$$\frac{\partial^2 u}{\partial t^2} = c^2\frac{\partial^2 u}{\partial x^2} \qquad \text{in } 0<x<l,\ t>0$$

with

$$u(x,0) = \chi(x), \qquad \frac{\partial u}{\partial t}(x,0) = 0,$$

$$u(0,t) = g(t), \qquad u(l,t) = 0.$$

As in E4.2.1, the Laplace transform $\bar{u}(x,p)$ satisfies

$$p^2\bar{u} - p\chi(x) = c^2\frac{\partial^2\bar{u}}{\partial x^2},$$

but now there are two boundary conditions

$$\bar{u}(0,p) = \bar{g}(p), \qquad \bar{u}(l,p) = 0.$$

The solution of this problem, obtained as in E2.1.5, may be expressed in the form

$$\bar{u}(x,p) = \bar{u}_1(x,p) + \bar{u}_2(x,p),$$

where

$$\bar{u}_1(x,p) = \bar{g}(p)\frac{\sinh[p(l-x)/c]}{\sinh(pl/c)}$$

and

$$\bar{u}_2(x,p) = \frac{\sinh(px/c)}{c\sinh(pl/c)}\int_0^l \chi(y)\sinh\left(\frac{p(l-y)}{c}\right)\mathrm{d}y$$
$$-\frac{1}{c}\int_0^x \chi(y)\sinh\left(\frac{p(x-y)}{c}\right)\mathrm{d}y$$
$$=\frac{\sinh[p(l-x)/c]}{c\sinh(pl/c)}\int_0^x \chi(y)\sinh\left(\frac{py}{c}\right)\mathrm{d}y$$
$$+\frac{\sinh(px/c)}{c\sinh(pl/c)}\int_x^l \chi(y)\sinh\left(\frac{p(l-y)}{c}\right)\mathrm{d}y.$$

To invert $\bar{u}_1(x,p)$, we write

$$\frac{\sinh[p(l-x)/c]}{\sinh(pl/c)} = \frac{\mathrm{e}^{p(l-x)/c}-\mathrm{e}^{-p(l-x)/c}}{\mathrm{e}^{pl/c}-\mathrm{e}^{-pl/c}}$$
$$=(\mathrm{e}^{-px/c}-\mathrm{e}^{-p(2l-x)/c})\sum_{n=0}^{\infty}\mathrm{e}^{-2npl/c},$$

so that

$$u_1(x,t) = \sum_{n=0}^{\infty}\left[g\left(t-\frac{x+2nl}{c}\right)\mathrm{H}\left(t-\frac{x+2nl}{c}\right)\right.$$
$$\left.-g\left(t-\frac{(2n+2)l-x}{c}\right)\mathrm{H}\left(t-\frac{(2n+2)l-x}{c}\right)\right].$$

Note that for any given values of $x$ and $t$ this is a finite series.

The function $\bar{u}_2(x,p)$ is regular except for simple poles where $\sinh(pl/c)=0$, that is at $p=n\pi\mathrm{i}c/l$ where $n$ is an integer. It is actually regular at $p=0$. The residue at $p=n\pi\mathrm{i}c/l$ is obtained from the first form as

$$\frac{\sinh(n\pi\mathrm{i}x/l)}{l\cosh n\pi\mathrm{i}}\int_0^l \chi(y)\sinh\left(\frac{n\pi\mathrm{i}(l-y)}{l}\right)\mathrm{d}y$$
$$=\frac{1}{l}\sin\left(\frac{n\pi x}{l}\right)\int_0^l \chi(y)\sin\left(\frac{n\pi y}{l}\right)\mathrm{d}y.$$

Hence the sum of the residues of $\bar{u}_2(x,p)\mathrm{e}^{pt}$ to the left of the path of integration in the inversion theorem is, on combining the terms $\pm n$,

$$u_2(x,t) = \sum_{n=1}^{\infty}\frac{2}{l}\int_0^l \chi(y)\sin\left(\frac{n\pi y}{l}\right)\mathrm{d}y\,\sin\left(\frac{n\pi x}{l}\right)\cos\left(\frac{n\pi ct}{l}\right).$$

We can also obtain a result analogous to that for $u_1(x, t)$ by writing the second form of $\bar{u}_2(x, p)$ as

$$\bar{u}_2(x, p)(1 - e^{-2pl/c})$$
$$= \frac{2}{c} e^{-pl/c}\left[\int_0^x \chi(y)\sinh\left(\frac{p(l-x)}{c}\right)\sinh\left(\frac{py}{c}\right)dy + \int_x^l \chi(y)\sinh\left(\frac{px}{c}\right)\sinh\left(\frac{p(l-y)}{c}\right)dy\right],$$

and expressing the hyperbolic functions in terms of exponentials. After some reduction, it is found that

$$\bar{u}_2(x, p)(1 - e^{-2pl/c}) = \tfrac{1}{2}[\bar{v}(x, p) + \bar{w}(x, p)],$$

where

$$v(x, t) = \begin{cases} \chi(x - ct) & (0 < t < x/c) \\ \chi(ct - x) & (x/c < t < (l + x)/c) \\ -\chi(2l + x - ct) & ((l + x)/c < t < 2l/c) \\ 0 & (t > 2l/c) \end{cases}$$

$$w(x, t) = \begin{cases} \chi(x + ct) & (0 < t < (l - x)/c) \\ -\chi(2l - x - ct) & ((l - x)/c < t < (2l - x)/c) \\ -\chi(x + ct - 2l) & ((2l - x)/c < t < 2l/c) \\ 0 & (t > 2l/c) \end{cases}$$

Thus $u_2(x, t)$ is the periodic function, with period $2l/c$, defined by

$$u_2(x, t) = \tfrac{1}{2}[v(x, t) + w(x, t)] \qquad (0 < t < 2l/c).$$

It remains to justify the evaluation of $u_2(x, t)$ as a sum of residues. Consider the rectangle with corners $a \pm (N + \frac{1}{2})\pi ic/l$, $-X \pm (N + \frac{1}{2})\pi ic/l$, where $a$ and $X$ are positive and $N$ is a positive integer. It will be proved that $|\bar{u}_2(x, p)| \to 0$ uniformly on the top and bottom of the rectangle as $N \to \infty$, and on the left side as $X \to \infty$.

Let $p = \xi + i\eta$. Since $\bar{u}_2(x, p)$ is an odd function of $p$, we can infer its behaviour in $\xi \leqslant 0$ from that in $\xi \geqslant 0$. The functions $\bar{v}(-p), \bar{w}(p)$ are Laplace transforms of functions $v(t)\, w(t)$ that vanish for $t > 2l/c$, so that (P3.2.6) $|\bar{v} + \bar{w}| \to 0$ as $\eta \to \pm\infty$ uniformly in $\xi \geqslant 0$, and as $\xi \to \infty$ uniformly in $\eta$. Also when $\xi \geqslant 0$

$$|1 - e^{-2pl/c}| \geqslant \begin{cases} 1 - e^{-2\xi l/c} & \text{for all } \eta, \\ 1 & \text{when } \eta = \pm(N + \frac{1}{2})\pi c/l. \end{cases}$$

Hence if we integrate along the line $\xi = a$ in the inversion theorem, we can apply P3.3.1 to express $u_2(x, t)$ as the sum of the residues.

The wave equation can also be treated in more than one space variable. In the following example, of supersonic flow past a slender body of revolution (Ward, 1955, Section 9.2), time is replaced by one of the space variables.

**E4.2.3** For small disturbances to a uniform supersonic stream with speed $U$ in the $x$ direction, the perturbation to the velocity field is given by $\mathbf{v} = \operatorname{grad} \phi$, where the potential $\phi$ satisfies

$$\nabla^2 \phi - M^2 \frac{\partial^2 \phi}{\partial x^2} = 0$$

and $M\,(>1)$ is the Mach number of the uniform stream. Consider the disturbance produced by a slender body of revolution with axis parallel to the stream. Let $(r, \theta, x)$ by cylindrical polar co-ordinates. Then for axisymmetric flow

$$\frac{\partial^2 \phi}{\partial r^2} + \frac{1}{r}\frac{\partial \phi}{\partial r} = B^2 \frac{\partial^2 \phi}{\partial x^2},$$

where $B = (M^2 - 1)^{1/2}$. If the body is given by $r = R(x)$ for $x > 0$ the boundary condition is

$$\frac{\partial \phi}{\partial x} = U \frac{\mathrm{d}R}{\mathrm{d}x} \qquad \text{at } r = R(x).$$

The Laplace transform

$$\bar{\phi}(r, p) = \int_0^\infty \phi(r, x) \mathrm{e}^{-px}\, \mathrm{d}x$$

satisfies

$$\frac{\partial^2 \bar{\phi}}{\partial r^2} + \frac{1}{r}\frac{\partial \bar{\phi}}{\partial r} = B^2 p^2 \bar{\phi},$$

since there is no disturbance at $x = 0$. The general solution of this equation is

$$\bar{\phi}(r, p) = C(p)\mathrm{I}_0(Bpr) + D(p)\mathrm{K}_0(Bpr).$$

The function $I_0(Bpr)$ is exponentially large as $r \to \infty$, so that (see E3.5.17) we must take $C(p) = 0$. Hence

$$\frac{\partial \bar{\phi}}{\partial r} = D(p)Bp\mathrm{K}_0'(Bpr) \sim -D(p)r^{-1} \qquad \text{as } r \to 0.$$

Because $R(x)$ is small, the boundary condition on $\phi$ can be approximated by

$$r\frac{\partial\phi}{\partial r} \to UR\frac{\mathrm{d}R}{\mathrm{d}x} \qquad \text{as } r \to 0.$$

In terms of the cross-sectional area $S(x) = \pi R^2(x)$, we therefore have

$$D(p) = -(U/2\pi)\mathscr{L}\{S'(x)\} = -(U/2\pi)p\bar{S}(p),$$

assuming that $S(0) = 0$. Consequently

$$\bar{\phi}(r, p) = -\frac{U}{2\pi}p\bar{S}(p)\mathrm{K}_0(Bpr).$$

The formula (Watson, 1944, Section 6.15 or 6.22)

$$\mathrm{K}_0(z) = \int_1^\infty (t^2 - 1)^{-1/2}\mathrm{e}^{-zt}\,\mathrm{d}t$$

now gives

$$\phi(r, x) = -\frac{U}{2\pi}S'(x)*[(x^2 - B^2r^2)^{-1/2}\mathrm{H}(x - Br)]$$

$$= -\frac{U}{2\pi}\int_0^{x-Br}\frac{S'(y)\,\mathrm{d}y}{[(x-y)^2 - B^2r^2]^{1/2}}\mathrm{H}(x - Br).$$

## 4.3 The Diffusion or Heat-Conduction Equation

The equation $\partial u/\partial t = \kappa\nabla^2 u$ governs mass transport by diffusion and heat transport by conduction. Here $\kappa$ is the diffusivity or thermometric conductivity, and $u$ denotes the concentration of the diffusing substance or the temperature of the conducting medium, usually measured relative to some reference temperature. The one- and two-dimensional forms of the equation also appear in problems of viscous flow.

For convenience, we shall follow Carslaw and Jaeger (1947) by writing $q = (p/\kappa)^{1/2}$ throughout this section. The first example is the problem of a semi-infinite medium ($x > 0$) in which the temperature (or concentration) is prescribed at time $t = 0$ for all $x > 0$ and at the boundary $x = 0$ when $t > 0$.

**E4.3.1**

$$\frac{\partial u}{\partial t} = \kappa \frac{\partial^2 u}{\partial x^2} \qquad \text{in } x > 0, t > 0$$

with

$$u(x, 0) = \chi(x) \qquad u(0, t) = g(t).$$

The Laplace transform $\bar{u}(x, p)$ satisfies

$$p\bar{u}(x, p) - \chi(x) = \kappa \frac{\partial^2 \bar{u}}{\partial x^2}, \qquad \bar{u}(0, p) = \bar{g}(p).$$

As with the wave equation in E4.2.1, we need to introduce a boundary condition as $x \to \infty$. The transformed equation has the general solution

$$\bar{u}(x, p) = A(p)e^{qx} + B(p)e^{-qx} - (\kappa q)^{-1}\chi(x) * \sinh qx,$$

where $A(p)$ and $B(p)$ are functions to be determined. This can be expressed as

$$\bar{u}(x, p) = \left( A(p) - \tfrac{1}{2}(\kappa q)^{-1}\bar{\chi}(q) + \tfrac{1}{2}(\kappa q)^{-1} \int_x^\infty \chi(y)e^{-qy}\,dy \right)e^{qx}$$
$$+ \left( B(p) + \tfrac{1}{2}(\kappa q)^{-1} \int_0^x \chi(y)e^{qy}\,dy \right)e^{-qx}.$$

In many problems it may be argued that $u(x, t)$, and hence $\bar{u}(x, p)$, must remain bounded as $x \to \infty$, so that

$$A(p) = \tfrac{1}{2}(\kappa q)^{-1}\bar{\chi}(q):$$

see also the note following this example. The condition at $x = 0$ then gives

$$B(p) + \tfrac{1}{2}(\kappa q)^{-1}\bar{\chi}(q) = \bar{g}(p),$$

so that

$$\bar{u}(x, p) = \tfrac{1}{2}(\kappa q)^{-1}\left( \int_x^\infty \chi(y)e^{-q(y-x)}\,dy + \int_0^x \chi(y)e^{-q(x-y)}\,dy \right.$$
$$\left. - \int_0^\infty \chi(y)e^{-q(x+y)}\,dy \right) + \bar{g}(p)e^{-qx}.$$

Now from E1.3.10 or E3.3.4

$$\mathscr{L}^{-1}\{\tfrac{1}{2}(\kappa q)^{-1}e^{-qx}\} = \theta(x, t) \equiv \tfrac{1}{2}(\pi\kappa t)^{-1/2}\exp\left( -\frac{x^2}{4\kappa t} \right),$$

$$\mathscr{L}^{-1}\{e^{-qx}\} = \phi(x, t) \equiv \tfrac{1}{2}x(\pi\kappa t^3)^{-1/2}\exp\left( -\frac{x^2}{4\kappa t} \right).$$

The function $\theta(x, t)$ represents an instantaneous plane source of heat, liberated on $x = 0$ at $t = 0$. Since

$$\phi(x, t) = -2\kappa \frac{\partial \theta}{\partial x}$$

it represents a plane of double sources on $x = 0$ at $t = 0$.

The inverse transform of $\bar{u}(x, p)$ is therefore

$$\begin{aligned} u(x, t) &= \int_x^\infty \chi(y)\theta(y - x, t)\,\mathrm{d}y + \int_0^x \chi(y)\theta(x - y, t)\,\mathrm{d}y \\ &\quad - \int_0^\infty \chi(y)\theta(y + x, t)\,\mathrm{d}y + g(t) * \phi(x, t) \\ &= \int_0^\infty \chi(y)[\theta(y - x, t) - \theta(y + x, t)]\,\mathrm{d}y + g(t) * \phi(x, t), \end{aligned} \tag{4.3.1}$$

since $\theta(-x, t) = \theta(x, t)$.

If we take $\chi(x) = \delta(x - x_1)$, we obtain a source at $x = x_1, t = 0$ together with a negative image source at $x = -x_1, t = 0$: the first term of $u(x, t)$ therefore represents an $x$-distribution of these sources. Similarly, the second term represents a $t$-distribution of double sources at $x = 0$ for $t > 0$.

Let us now consider two special cases.

(i) $\chi(x) = T_0[\mathrm{H}(x) - \mathrm{H}(x - a)], \quad g(t) = 0.$

Here the region $0 < x < a$ is raised to the initial temperature $T_0$ above the ambient temperature, which is maintained unaltered at $x = 0$. The solution for $\bar{u}(x, p)$ is

$$\bar{u}(x, p) = \begin{cases} \frac{1}{2}T_0 p^{-1}[2 - 2\mathrm{e}^{-qx} - \mathrm{e}^{-q(a-x)} + \mathrm{e}^{-q(a+x)}] & (0 < x < a), \\ \frac{1}{2}T_0 p^{-1}[-2\mathrm{e}^{-qx} + \mathrm{e}^{-q(x-a)} + \mathrm{e}^{-q(x+a)}] & (x > a). \end{cases}$$

The inverse transform (E3.3.4)

$$\mathscr{L}^{-1}\{p^{-1}\exp(-\alpha\sqrt{p})\} = \operatorname{erfc}\left(\frac{\alpha}{2\sqrt{t}}\right)$$

gives the solution in the form

$$u(x, t) = \begin{cases} T_0[1 - \operatorname{erfc}(\hat{x}) - \frac{1}{2}\operatorname{erfc}(\hat{a} - \hat{x}) \\ \quad + \frac{1}{2}\operatorname{erfc}(\hat{a} + \hat{x})] & (0 < x < a), \\ T_0[-\operatorname{erfc}(\hat{x}) + \frac{1}{2}\operatorname{erfc}(\hat{x} - \hat{a}) \\ \quad + \frac{1}{2}\operatorname{erfc}(\hat{x} + \hat{a})] & (x > a), \end{cases}$$

where $\hat{x} = x/\sqrt{(4\kappa t)}$ and $\hat{a} = a/\sqrt{(4\kappa t)}$. The general solution (4.3.1) leads to the more symmetrical form

$$u(x, t) = \tfrac{1}{2} T_0 [\operatorname{erfc}(\hat{a} + \hat{x}) - \operatorname{erfc}(\hat{x}) - \operatorname{erfc}(\hat{a} - \hat{x}) + \operatorname{erfc}(-\hat{x})],$$

which is equivalent because $\operatorname{erf}(\xi) = 1 - \operatorname{erfc}(\xi)$ is an odd function of $\xi$.

(ii) $\chi(x) = 0 \qquad g(t) = T_0 \cos(\omega t + \alpha)$.
Here the medium is initially at zero temperature and the face $x = 0$ is given a sinusoidally varying temperature. For this case,

$$\bar{u}(x, p) = T_0 \frac{p \cos\alpha - \omega \sin\alpha}{p^2 + \omega^2} e^{-qx},$$

which has simple poles at $p = \pm i\omega$ and a branch point at $p = 0$. When we apply Jordan's lemma to the inverse transform, we get $u(x, t)$ as the sum of the residues of $u(x, t)$ at the poles together with a loop integral round a cut along the negative real axis. This loop can be made to coincide with the two edges of the cut, on which we put $p = \kappa r e^{\pm i\pi}$. In this way, we obtain

$$u(x, t) = T_0 \exp[-\tfrac{1}{2}\sqrt{(\omega/\kappa)}x] \cos[\omega t + \alpha - \tfrac{1}{2}\sqrt{(\omega/\kappa)}x]$$
$$- \frac{\kappa T_0}{\pi} \int_0^\infty \frac{\kappa r \cos\alpha + \omega \sin\alpha}{\kappa^2 r^2 + \omega^2} \sin(x\sqrt{r}) e^{-\kappa t r} \, dr. \qquad (4.3.2)$$

It is more useful to examine the behaviour of $u(x, t)$ for large and small values of $t$. When $t \to \infty$, the asymptotic expansion of the loop integral is found as in P3.4.5 by expanding $\bar{u}(x, p)$ about the branch point $p = 0$ or, what is equivalent, by applying P1.2.13 to the integral in (4.3.2). Hence as $t \to \infty$ with $x$ fixed

$$u(x, t) = T_0 \exp[-\tfrac{1}{2}\sqrt{(\omega/\kappa)}x] \cos[\omega t + \alpha - \tfrac{1}{2}\sqrt{(\omega/\kappa)}x]$$
$$- \frac{T_0}{\sqrt{(\pi\kappa t)}} \left( \frac{x \sin\alpha}{2\omega t} + \frac{3\kappa x \cos\alpha - \tfrac{1}{2}\omega x^3 \sin\alpha}{4\kappa\omega^2 t^2} + O(t^{-3}) \right).$$

The integral is thus the transient part of the solution, though it decays only like $t^{-3/2}$.

For $t \to 0$, we expand for large $p$ as

$$\bar{u}(x, p) = \frac{T_0}{p} \left( \cos\alpha - \frac{\omega}{p} \sin\alpha \right) \sum_{n=0}^{\infty} \left( -\frac{\omega^2}{p^2} \right)^n e^{-qx}.$$

Since (E3.3.4)

$$\mathscr{L}^{-1}\{p^{-m-1} \exp(-\lambda\sqrt{p})\} = (4t)^m i^{2m} \operatorname{erfc}\left( \frac{\lambda}{2\sqrt{t}} \right),$$

where $\mathrm{i}^m \operatorname{erfc}$ is an integrated error function, we have

$$u(x,t) = T_0 \sum_{n=0}^{\infty} (-\omega^2)^n \left[ 2^{4n} t^{2n} \cos\alpha\, \mathrm{i}^{4n} \operatorname{erfc}\left( \frac{x}{2\sqrt{(\kappa t)}} \right) - 2^{4n+2} t^{2n+1} \sin\alpha\, \mathrm{i}^{4n+2} \operatorname{erfc}\left( \frac{x}{2\sqrt{(\kappa t)}} \right) \right].$$

This series converges for all $t$.

**Note** If $\bar{v}(x,p) = A(p)\mathrm{e}^{qx}$ is a Laplace transform for all $x \geqslant 0$, we have $\bar{v}(0,p) = \bar{v}(x,p)\mathrm{e}^{-qx}$ for any $x > 0$. Hence

$$v(0,t) = \int_0^t v(x,s)\phi(x,t-s)\,\mathrm{d}s.$$

As a function of $t$, $\phi(x,t)$ increases with $t$ when $0 < 6\kappa t < x^2$. Hence

$$|v(0,t)| \leqslant \int_0^t |v(x,s)|\,\mathrm{d}s\,\phi(x,t)$$

for all $x \geqslant (6\kappa t)^{1/2}$. Consequently, if we assume merely that

$$|v(x,t)| < F(t)\exp(\delta x^2) \qquad \text{for all } \delta > 0,$$

where $F(t)$ is integrable, then we must have $v(0,t) \equiv 0$ and hence $A(p) = 0$.

The case of the finite slab follows a similar line to E4.2.2.

**E4.3.2**

$$\frac{\partial u}{\partial t} = \kappa \frac{\partial^2 u}{\partial x^2} \qquad \text{in } 0 < x < l,\ t > 0$$

with

$$u(x,0) = \chi(x), \qquad u(0,t) = g(t), \qquad u(l,t) = 0.$$

The solution of the transformed equation is (as in E4.2.2)

$$\bar{u}(x,p) = \bar{u}_1(x,p) + \bar{u}_2(x,p),$$

where

$$\bar{u}_1(x,p) = \bar{g}(p)\frac{\sinh[q(l-x)]}{\sinh ql}$$

and

$$\bar{u}_2(x,p) = \frac{\sinh qx}{\kappa q \sinh ql}\int_0^l \chi(y)\sinh[q(l-y)]\,\mathrm{d}y$$

$$- \int_0^x \frac{\chi(y)}{\kappa q}\sinh\,[q(x-y)]\,\mathrm{d}y.$$

$$= \int_0^x \chi(y)\frac{\sinh qy \sinh[q(l-x)]}{\kappa q \sinh ql}\,\mathrm{d}y$$

$$+ \int_x^l \chi(y)\frac{\sinh[q(l-y)]\sinh qx}{\kappa q \sinh ql}\,\mathrm{d}y.$$

This time $\sinh[q(l-x)]/\sinh ql$ is a Laplace transform. It is regular at $p=0$ and has simple poles where $ql = n\pi\mathrm{i}$, that is $p = -n^2\pi^2\kappa/l^2$, for $n = 1, 2, 3, \ldots$ By calculating the residues at these poles, we obtain

$$\mathscr{L}^{-1}\left\{\frac{\sinh[q(l-x)]}{\sinh ql}\right\} = \sum_{n=1}^{\infty}\frac{2n\pi x}{l^2}\sin\left(\frac{n\pi x}{l}\right)\exp(-n^2\pi^2\kappa t/l^2).$$

We can also treat the inverse transform by a similar method to that used in E4.2.2, which gives

$$\mathscr{L}^{-1}\left\{\frac{\sinh[q(l-x)]}{\sinh ql}\right\} = \mathscr{L}^{-1}\left\{(\mathrm{e}^{-qx} - \mathrm{e}^{-q(2l-x)})\sum_{n=0}^{\infty}\mathrm{e}^{-2nql}\right\}$$

$$= \sum_{n=0}^{\infty}[\phi(x+2nl,t) - \phi((2n+2)l-x,t)]$$

$$= \sum_{n=-\infty}^{\infty}\phi(x+2nl,t).$$

The first form converges rapidly unless $t$ is very small; so does the second unless $t$ is quite large. The inverse transform can also be expressed as

$$\mathscr{L}^{-1}\left\{\frac{\sinh[q(l-x)]}{\sinh ql}\right\} = -\kappa\frac{\partial}{\partial x}\vartheta_3\left(\frac{x}{2l}\middle|\frac{\mathrm{i}\pi\kappa t}{l^2}\right),$$

where the theta function is defined in D3.5.1. This theta function represents an instantaneous source of heat at $x=0, t=0$ with insulating walls at $x = l, -l$. Consequently

$$u_1(x,t) = g(t) * \left[-\kappa\frac{\partial}{\partial x}\vartheta_3\left(\frac{x}{2l}\middle|\frac{\mathrm{i}\pi\kappa t}{l^2}\right)\right]$$

represents a time distribution of double sources on the wall $x = 0$.

The use of P3.3.1 to calculate $u_1(x, t)$ can be justified as in E3.3.2 by considering integrals along the parabolae given by

$$ql = (N + \tfrac{1}{2})\pi i + \xi.$$

This applies equally to the second form of $\bar{u}_2(x, p)$. The residues are more easily found from the first form, which gives

$$u_2(x,t) = \sum_{n=1}^{\infty} \frac{2}{l} \int_0^l \chi(y) \sin\left(\frac{n\pi y}{l}\right) \mathrm{d}y \sin\left(\frac{n\pi x}{l}\right) \exp(-n^2\pi^2\kappa t/l^2)$$

$$= \frac{1}{l} \int_0^l \chi(y) \left[ \vartheta_3\left(\frac{y-x}{2l} \middle| \frac{i\pi\kappa t}{l^2}\right) - \vartheta_3\left(\frac{y+x}{2l} \middle| \frac{i\pi\kappa t}{l^2}\right)\right] \mathrm{d}y.$$

The expansion method for small $t$ gives

$$u_2(x,t) = \int_0^l \frac{\chi(y)}{4\sqrt{(\pi\kappa t)}} \sum_{n=-\infty}^{\infty} [\theta(2nl + x - y, t) - \theta(2nl + x + y, t)] \mathrm{d}y.$$

The two forms for $u_2(x, t)$ are related by the transformation of the theta function given in D3.5.1.

The last problem of this section is a more difficult one, that of the flow of heat in a circular cylinder which loses heat from its surface at a rate proportional to the excess temperature.

**E4.3.3**

$$\frac{\partial u}{\partial t} = \kappa\left(\frac{\partial^2 u}{\partial r^2} + \frac{1}{r}\frac{\partial u}{\partial r}\right) \qquad \text{in } 0 < r < a,\ t > 0$$

with

$$u(r, 0) = \chi(r) \qquad \text{and} \qquad \frac{\partial u}{\partial r} = -hu \qquad \text{at } r = a.$$

Here $h(\geqslant 0)$ is a constant.

The transformed equation

$$p\bar{u} - \chi(r) = \kappa\left(\frac{\partial^2 \bar{u}}{\partial r^2} + \frac{1}{r}\frac{\partial \bar{u}}{\partial r}\right)$$

has complementary functions $\mathrm{I}_0(qr)$ and $\mathrm{K}_0(qr)$. Solving by the method of variation of parameters, we find that

$$\kappa\bar{u}(r,p) = \left(A(p) - \int_0^r s\chi(s)\mathrm{K}_0(qs)\mathrm{d}s\right)\mathrm{I}_0(qr)$$

$$+ \left(B(p) + \int_0^r s\chi(s)\mathrm{I}_0(qs)\mathrm{d}s\right)\mathrm{K}_0(qr),$$

where the Wronskian relation (Watson, 1944, Section 3.71)

$$\mathrm{I}_0(z)\mathrm{K}_0'(z) - \mathrm{K}_0(z)\mathrm{I}_0'(z) = -z^{-1}$$

has been used.

We require that $u(r, t)$, and hence $\bar{u}(r, p)$, shall remain finite as $r \to 0$, so that $B(p) = 0$. The condition at $r = a$, $\partial\bar{u}/\partial r = -h\bar{u}$, now gives

$$\left(A(p) - \int_0^a s\chi(s)\mathrm{K}_0(qs)\mathrm{d}s\right)[q\mathrm{I}_0'(qa) + h\mathrm{I}_0(qa)]$$

$$+ \int_0^a s\chi(s)\mathrm{I}_0(qs)\mathrm{d}s[q\mathrm{K}_0'(qa) + h\mathrm{K}_0(qa)] = 0.$$

Hence

$$\bar{u}(r, p) = \int_0^a s\chi(s)\bar{v}(r, s, p)\mathrm{d}s,$$

where

$$\bar{v}(r, s, p) = -\frac{q\mathrm{K}_0'(qa) + h\mathrm{K}_0(qa)}{q\mathrm{I}_0'(qa) + h\mathrm{I}_0(qa)}\mathrm{I}_0(qr)\mathrm{I}_0(qs) + \begin{cases} \mathrm{K}_0(qs)\mathrm{I}_0(qr) & \text{for } r < s, \\ \mathrm{K}_0(qr)\mathrm{I}_0(qs) & \text{for } r > s. \end{cases}$$

Following the method of E4.3.2, we now examine the singularities of the function $\bar{v}(r, s, p)$. Since $\mathrm{I}_0(z)$ and $[\mathrm{K}_0(z) + \mathrm{I}_0(z)\log z]$ are both even regular functions of $z$, it can be shown that if $h > 0$ $\bar{v}(r, s, p)$ is regular at $p = 0$, whereas if $h = 0$ it has a simple pole with residue $2/a^2$. There are also poles where $q\mathrm{I}_0'(qa) + h\mathrm{I}_0(qa) = 0$. If $qa = \mathrm{i}\alpha$, this equation becomes

$$\alpha\mathrm{J}_0'(\alpha) + ha\mathrm{J}_0(\alpha) = 0. \tag{4.3.3}$$

It is known (Watson, 1944, Sections 15.23, 15.25) that equation (4.3.3) has an infinity of roots, and that when $ha \geqslant 0$ these roots are all real and simple (except $\alpha = 0$ when $ha = 0$). Thus $\bar{v}(r, s, p)$ has a corresponding set of poles on the negative real axis. The residue at the pole $p = -\alpha^2\kappa/a^2$ is

$$-\frac{\mathrm{i}\alpha\mathrm{K}_0'(\mathrm{i}\alpha) + ha\mathrm{K}_0(\mathrm{i}\alpha)}{[\mathrm{d}\{qa\mathrm{I}_0'(qa) + ha\mathrm{I}_0(qa)\}/\mathrm{d}q]_{q=\mathrm{i}\alpha}} 2\frac{\mathrm{i}\alpha}{a}\mathrm{I}_0\left(\frac{\mathrm{i}\alpha r}{a}\right)\mathrm{I}_0\left(\frac{\mathrm{i}\alpha s}{a}\right),$$

which simplifies, with the aid of the Wronskian relation, to

$$\frac{2\alpha^2 J_0(\alpha r/a) J_0(\alpha s/a)}{a^2(\alpha^2 + h^2 a^2) J_0^2(\alpha)}.$$

To justify the use of P3.3.1, we have to show that $|\bar{v}| \to 0$ on a suitable set of contours. The asymptotic form for $J_0(z)$ (E3.4.3) shows that when $\alpha$ is large it is given approximately by $\cos(\alpha - \frac{3}{4}\pi) = 0$. Hence we choose the parabolae given by $qa = (N + \frac{3}{4})\pi i + \xi$, and then use the asymptotic forms of $I_0(z)$ and $K_0(z)$ (Watson, 1944, Section 7.23) to show that $|\bar{v}| \to 0$ uniformly as $N \to \infty$.

Thus if $\alpha_1, \alpha_2, \ldots$ are the positive roots of equation (4.3.3), the solution when $h > 0$ is

$$u(r, t) = \sum_{n=1}^{\infty} c_n J_0\left(\alpha_n \frac{r}{a}\right) \exp\left(-\alpha_n^2 \frac{\kappa t}{a^2}\right),$$

where

$$c_n = \frac{2\alpha_n^2}{a^2(\alpha_n^2 + h^2 a^2) J_0^2(\alpha_n)} \int_0^a s\chi(s) J_0\left(\alpha_n \frac{s}{a}\right) ds.$$

For $h = 0$, which corresponds to an insulated surface,

$$u(r, t) = c_0 + \sum_{n=1}^{\infty} c_n J_0\left(\alpha_n \frac{r}{a}\right) \exp\left(-\alpha_n^2 \frac{\kappa t}{a^2}\right),$$

where

$$c_0 = \frac{2}{a^2} \int_0^a s\chi(s) ds,$$

$$c_n = \frac{2}{a^2 J_0^2(\alpha_n)} \int_0^a s\chi(s) J_0\left(\alpha_n \frac{s}{a}\right) ds,$$

and $\alpha_1, \alpha_2, \ldots$ are now the positive roots of $J_0'(\alpha) = 0$.

As with E4.3.2, this solution converges very rapidly unless $t$ is quite small. Asymptotic results for $t \to 0$ are found by expanding $\bar{v}$ for large $q$. The asymptotic forms of the Bessel functions give, after some reduction,

$$\bar{v}(r, s, p) = \frac{e^{-(2a-r-s)q}}{2\kappa q\sqrt{(rs)}}\left(1 + \frac{2(3 - 8ha)a^{-1} + r^{-1} + s^{-1}}{8q} + O(q^{-2})\right)$$
$$+ \frac{e^{-|r-s|q}}{2\kappa q\sqrt{(rs)}}\left(1 + \frac{|r^{-1} - s^{-1}|}{8q} + O(q^{-2})\right),$$

from which

$$v(r,s,t)=\frac{1}{\sqrt{(rs)}}\left\{\frac{1}{2\sqrt{(\pi\kappa t)}}\exp\left(-\frac{(2a-r-s)^2}{4\kappa t}\right)\right.$$

$$+\frac{1}{16}\left[\frac{2}{a}(3-8ha)+\frac{1}{r}+\frac{1}{s}\right]\operatorname{erfc}\left(\frac{2a-r-s}{2\sqrt{(\kappa t)}}\right)+\dots$$

$$+\frac{1}{2\sqrt{(\pi\kappa t)}}\exp\left(-\frac{(r-s)^2}{4\kappa t}\right)$$

$$\left.+\frac{1}{16}\left|\frac{1}{r}-\frac{1}{s}\right|\operatorname{erfc}\left(\frac{|r-s|}{2\sqrt{(\kappa t)}}\right)+\dots\right\}.$$

When $t$ is small, the first group of terms is very small unless $r$ and $s$ are both nearly equal to $a$, so that they represent the effect of the boundary: the second group is very small unless $r$ and $s$ are nearly equal so that these terms give local diffusion. This result fails when $rs$ is small. The first group is then exponentially small as $t\to 0$, so that we need consider only the local diffusion effect. This is given by

$$v(r,s,t)=\begin{cases}\mathscr{L}^{-1}\{\kappa^{-1}\mathrm{K}_0(qs)\mathrm{I}_0(qr)\} & (r<s),\\ \mathscr{L}^{-1}\{\kappa^{-1}\mathrm{K}_0(qr)\mathrm{I}_0(qs)\} & (r>s),\end{cases}$$

$$=\frac{1}{2\kappa t}\exp\left(-\frac{r^2+s^2}{4\kappa t}\right)\mathrm{I}_0\left(\frac{rs}{2\kappa t}\right)$$

(see E3.5.13).

In the particular case when $u(r,0)=T_0$ is a constant, we have

$$\bar{u}(r,p)=\frac{T_0}{p}\left(1-\frac{h\mathrm{I}_0(qr)}{q\mathrm{I}_0'(qa)+h\mathrm{I}_0(qa)}\right).$$

If $h=0$, the solution is simply $u(r,t)=T_0$. For $h>0$, we get by the residue method that

$$u(r,t)=\sum_{n=1}^{\infty}\frac{2T_0ha\,\mathrm{J}_0(\alpha_n r/a)}{(\alpha_n^2+h^2a^2)\mathrm{J}_0(\alpha_n)}\exp(-\alpha_n^2\kappa t/a^2),$$

where $\alpha_1,\alpha_2,\dots$ are the roots of equation (4.3.3).

For small $t$, we expand for large $q$ as

$$\bar{u}(r,p)=\frac{T_0}{p}\left\{1-h\left(\frac{a}{r}\right)^{1/2}\mathrm{e}^{-(a-r)q}\right.$$

$$\left.\times\left[\frac{1}{q}+\frac{a-(3-8ha)r}{8q^2ra}+\mathrm{O}\left(\frac{1}{q^3}\right)\right]\right\}$$

to obtain

$$u(r,t) = T_0\left\{1 - ha\left[2\left(\frac{\kappa t}{ar}\right)^{1/2}\operatorname{ierfc}\left(\frac{a-r}{2\sqrt{(\kappa t)}}\right)\right.\right.$$
$$\left.\left. + \frac{a-(3-8ha)r}{(ar)^{1/2}}\frac{2\kappa t}{ar}\mathrm{i}^2\operatorname{erfc}\left(\frac{a-r}{2\sqrt{(\kappa t)}}\right) + \ldots\right]\right\},$$

provided that $r$ is not small. In this case, there is no local diffusion, and the difficulty at $r = 0$ arises from the penetration of the boundary effect to the region near the axis. When $r$ is small, we expand $\bar{u}(r,p)$ on the assumption that $qa$ is large but $qr$ is small, so that

$$\bar{u}(r,p) = \frac{T_0}{p}\left[1 - \sqrt{(2\pi)}\frac{ha}{(qa)^{1/2}}\left(1 + \frac{\frac{3}{8} - ha}{qa} + \ldots\right)\right.$$
$$\left.\times\left[1 + (\tfrac{1}{2}qr)^2 + \ldots\right]\mathrm{e}^{-qa}\right].$$

From E3.3.4

$$\mathscr{L}^{-1}\{p^{-1}(qa)^{-1/2}\mathrm{e}^{-qa}\} = \frac{a}{\pi\sqrt{(2\kappa t)}}\exp\left(-\frac{a^2}{8\kappa t}\right)\mathrm{K}_{1/4}\left(\frac{a^2}{8\kappa t}\right),$$

and the other inverse transforms needed may be deduced from this. In this way we get when $r$ and $t$ are both small

$$u(r,t) = T_0 - \sqrt{\left(\frac{2}{\pi}\right)}haT_0\exp(-\tfrac{1}{4}\lambda^2)\left(\lambda(\mathrm{K}_{3/4} - \mathrm{K}_{1/4})\right.$$
$$+ (\tfrac{3}{8} - ha)\frac{2}{3\lambda}[(1-\lambda^2)\mathrm{K}_{1/4} - \lambda^2\mathrm{K}_{3/4}] + \ldots$$
$$\left. + \frac{r^2}{4a^2}[\tfrac{1}{2}\lambda^3(\mathrm{K}_{1/4} + \mathrm{K}_{3/4}) + (\tfrac{3}{8} - ha)\lambda\mathrm{K}_{1/4} + \ldots] + \ldots\right),$$

where $\lambda = a/\sqrt{(2\kappa t)}$ and the argument of each Bessel function is $\frac{1}{4}\lambda^2 = a^2/8\kappa t$.

## 4.4 Other Equations

In this section we shall treat some further partial differential equations by the Laplace transform method. Although these resemble the wave or diffusion equations, they present some new features of interest.

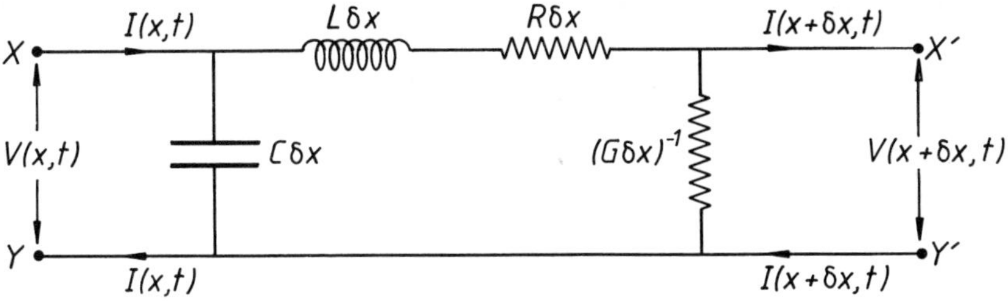


Figure 4.3

**E4.4.1** The transmission-line equations are obtained by representing a length $\delta x$ of the line by the circuit shown in Figure 4.3. Current $I(x, t)$ flows in at X and out at Y, and $I(x+\delta x, t)$ flows out at X′ and in at Y′. The potential difference is $V(x, t)$ between X and Y, and $V(x+\delta x, t)$ between X′ and Y′. In the limit as $\delta x \to 0$ the equations for this circuit give the partial differential equations

$$L\frac{\partial I}{\partial t} + RI + \frac{\partial V}{\partial x} = 0,$$

$$C\frac{\partial V}{\partial t} + GV + \frac{\partial I}{\partial x} = 0.$$

Here $L, R, C$ and $G$ are respectively the inductance, resistance, capacity and leakage conductance per unit length of the line. In general they may all vary with $x$, but in this example they will be taken as constants.

When $L$, $R$, $C$ and $G$ are constant, elimination of $I$ gives

$$\frac{\partial^2 V}{\partial x^2} - LC\frac{\partial^2 V}{\partial t^2} - (LG + RC)\frac{\partial V}{\partial t} - RGV = 0,$$

and similarly we find that $I$ satisfies the same equation. The characteristics of the equation are given by $LC(\mathrm{d}x)^2 - (\mathrm{d}t)^2 = 0$, so that if $LC \neq 0$ the equation is hyperbolic, with characteristics given by $(\mathrm{d}x/\mathrm{d}t)^2 = \pm c$, where $c = (LC)^{-1/2}$. [This is true whether or not $L$, $R$, $C$ and $G$ are constant.]

If $LC = 0$, the equation is parabolic and can be reduced to the heat conduction equation by writing $V(x, t) = u(x, t)\mathrm{e}^{-\alpha t}$, where $\alpha = G/C$ if $C \neq 0$ or $\alpha = R/L$ if $L \neq 0$. The hyperbolic case can be reduced to a standard form by putting

$$V(x, t) = v(x, t)\mathrm{e}^{-\beta t}, \qquad \text{where } \beta = \frac{LG + RC}{2LC}.$$

This gives

$$\frac{\partial^2 v}{\partial t^2} = c^2 \frac{\partial^2 v}{\partial x^2} + k^2 v, \qquad \text{where } k = \frac{|LG - RC|}{2LC}.$$

If $LG = RC$, the transmission line equations are reduced exactly to the wave equation. This case is *Heaviside's distortionless line*, in which signals are propagated without change of form, though attenuated by the factor $\mathrm{e}^{-\beta t}$.

Consider the problem of a semi-infinite transmission line which is initially dead, that is $V = I = 0$ at $t = 0$, with the potential difference $V = g(t)$ applied at the end $x = 0$. The initial conditions for $v(x, t)$ are $v = \partial v/\partial t = 0$ at $t = 0$, so that its transform satisfies

$$c^2 \frac{\partial^2 \bar{v}}{\partial x^2} - (p^2 - k^2)\bar{v} = 0. \tag{4.4.1}$$

Also $v(0, t) = g(t)\mathrm{e}^{\beta t}$, so that $\bar{v}(0, p) = \bar{g}(p - \beta)$. From (4.4.1)

$$\bar{v}(p) = A(p)\exp[(p^2 - k^2)^{1/2}\bar{x}/c] + B(p)\exp[-(p^2 - k^2)^{1/2}x/c],$$

but as in E4.2.1 we must reject the positive exponential. Hence

$$\bar{v}(x, p) = \bar{g}(p - \beta)\exp\left[-(p^2 - k^2)^{1/2}\frac{x}{c}\right].$$

Now when $|p| \to \infty$

$$\exp[-a\sqrt{(p^2 - k^2)}] = \mathrm{e}^{-ap}\left[1 + \frac{k^2 a}{2p} + \mathrm{O}(p^{-2})\right],$$

so that it is not a Laplace transform. We therefore consider

$$\bar{w}(a, p) = \exp[-a\sqrt{(p^2 - k^2)}] - \mathrm{e}^{-ap},$$

where $a > 0$. The inverse transform of $\bar{w}(a, p)$ may be evaluated by the method of E3.5.8, or deduced from that example by differentiation with respect to $a$. In this way, we find that

$$w(a, t) = \frac{ka}{\sqrt{(t^2 - a^2)}}\mathrm{I}_1(k\sqrt{(t^2 - a^2)})\mathrm{H}(t - a).$$

We now write

$$\bar{v}(x, p) = \bar{g}(p - \beta)\mathrm{e}^{-px/c} + \bar{g}(p - \beta)\bar{w}\left(\frac{x}{c}, p\right),$$

and obtain on inversion

$$v(x,t)$$
$$= g\left(t-\frac{x}{c}\right)\exp\left[\beta\left(t-\frac{x}{c}\right)\right]\mathrm{H}\left(t-\frac{x}{c}\right)+[g(t)\mathrm{e}^{\beta t}]*w\left(\frac{x}{c},t\right).$$

The potential difference is therefore

$$V(x,t)=\left[g\left(t-\frac{x}{c}\right)\mathrm{e}^{-\beta x/c}+\int_{x/c}^{t}\frac{kx}{\sqrt{(c^2s^2-x^2)}}\right.$$
$$\left.\times \mathrm{I}_1\left(k\sqrt{\left(s^2-\frac{x^2}{c^2}\right)}\right)g(t-s)\mathrm{e}^{-\beta s}\mathrm{d}s\right]\mathrm{H}\left(t-\frac{x}{c}\right).$$

The second term is zero when $k=0$ (the distortionless line) and when $t=x/c$. It therefore represents the distortion produced by the line after the arrival of the signal.

As special cases, let $V$ be either constant or sinusoidal at $x=0$. If $V(0,t)=V_0\mathrm{H}(t)$ we have

$$\bar{v}(x,p)=V_0(p-\beta)^{-1}\exp\left(-\frac{x}{c}\sqrt{(p^2-k^2)}\right),$$

so that

$$\bar{V}(x,p)=V_0p^{-1}\exp\left(-\frac{x}{c}\sqrt{[(p+\beta)^2-k^2]}\right).$$

The asymptotic form of $\bar{V}(x,p)$ as $p\to\infty$, namely

$$\bar{V}(x,p)=V_0p^{-1}\left(1+\frac{k^2x}{2cp}+\mathrm{O}(p^{-2})\right)\mathrm{e}^{-(p+\beta)x/c},$$

gives the behaviour of $V(x,t)$ when $(t-x/c)$ is small as

$$V(x,t)=V_0\mathrm{e}^{-\beta x/c}\left[1+\frac{1}{2}k^2\frac{x}{c}\left(t-\frac{x}{c}\right)+\mathrm{O}\left(t-\frac{x}{c}\right)^2\right]\mathrm{H}\left(t-\frac{x}{c}\right).$$

The form of $V(x,t)$ as $t\to\infty$ is found from the singularities of $\bar{V}(x,p)$, which are a pole at $p=0$ and branch points at $p=-\beta\pm k$, which are both real and negative. The residue at $p=0$ is

$$V_0\exp\left(-\frac{x}{c}\sqrt{(\beta^2-k^2}\right)=V_0\exp[-\sqrt{(RG)}x].$$

The branch point at $-\beta+k$ is the dominant one, so we put $p=-\beta+k+q$ and expand for small $q$ as a power series in $q^{1/2}$, giving

$$\bar{V}(x, p) = V_0\left(-\frac{1}{\beta - k} + \frac{\sqrt{(2k)}}{\beta - k}\frac{x}{c}q^{1/2} + \mathrm{O}(q)\right).$$

The first term is regular and so does not contribute. Hence

$$V(x, t) = V_0 \exp[-\sqrt{(RG)}x] - V_0 \exp[-(\beta - k)t]$$

$$\times\left[\sqrt{\left(\frac{k}{2\pi}\right)}\frac{x}{(\beta - k)c}t^{-3/2} + \mathrm{O}(t^{-5/2})\right].$$

Thus when $t = x/c$, $V$ jumps suddenly from 0 to

$$V_0 \exp(-\beta x/c) = V_0 \exp\left(-\frac{LG + RC}{2\sqrt{(LC)}}x\right)$$

and thereafter rises gradually, approaching its limiting value $V_0 \exp[-\sqrt{(RG)}x]$ with exponential rapidity as $t \to \infty$.

If $V(0, t) = V_0 \cos(\omega t + \alpha)\mathrm{H}(t)$ we find that

$$\bar{V}(x, p) = V_0 \frac{p\cos\alpha - \omega\sin\alpha}{p^2 + \omega^2}\exp\left(-\frac{x}{c}\sqrt{[(p + \beta)^2 - k^2]}\right).$$

As $p \to \infty$

$$\bar{V}(x, p) = V_0 \frac{p\cos\alpha - \omega\sin\alpha}{p^2 + \omega^2}\left(1 + \frac{k^2 x}{2cp} + \mathrm{O}(p^{-2})\right)\mathrm{e}^{-(p+\beta)x/c},$$

so that near $t = x/c$

$$V(x, t) = V_0 \mathrm{e}^{-\beta x/c}\left\{\cos\left[\omega\left(t - \frac{x}{c}\right) + \alpha\right]\right.$$

$$\left. + \frac{k^2 x\cos\alpha}{2c}\left(t - \frac{x}{c}\right) + \dots\right\}\mathrm{H}\left(t - \frac{x}{c}\right).$$

In this case, $\bar{V}(x, p)$ has poles at $p = \pm\mathrm{i}\omega$ as well as the branch points at $p = -\beta \pm k$. The poles contribute a steady oscillation. The residue of $\bar{V}(x, p)\mathrm{e}^{pt}$ at $p = \mathrm{i}\omega$ is

$$\tfrac{1}{2}V_0 \exp\left[\mathrm{i}(\omega t + \alpha) - (\beta + \mathrm{i}\omega - k)^{1/2}(\beta + \mathrm{i}\omega + k)^{1/2}\frac{x}{c}\right],$$

so that if we write

$$\beta + \mathrm{i}\omega - k = A_1 \exp(\mathrm{i}\theta_1), \qquad \beta + \mathrm{i}\omega + k = A_2 \exp(\mathrm{i}\theta_2),$$

where $\theta_1$ and $\theta_2$ are acute angles, the poles combine to give

$$V_0 \exp\left[-(A\cos\theta)\frac{x}{c}\right]\cos\left[\omega t+\alpha-(A\sin\theta)\frac{x}{c}\right],$$

where $A=\sqrt{(A_1A_2)}$ and $\theta=\frac{1}{2}(\theta_1+\theta_2)$. Near the branch point at $k-\beta$

$$\bar{V}(x,p)=V_0\left(\frac{(k-\beta)\cos\alpha-\omega\sin\alpha}{(k-\beta)^2+\omega^2}+O(q)\right)$$
$$\times\left(1-(2kq)^{1/2}\frac{x}{c}+O(q)\right),$$

where $p=k-\beta+q$, so that the leading term of the transient part is

$$-V_0\sqrt{\left(\frac{k}{2\pi}\right)}\frac{\cos(\theta_1-\alpha)}{A_1}\frac{x}{c}t^{-3/2}e^{-(\beta-k)t}.$$

**E4.4.2** The equation governing the flexural vibrations of a bar, neglecting rotatory inertia, is (Love, 1927, Section 280)

$$\frac{\partial^2 u}{\partial t^2}=-\kappa^2\frac{\partial^4 u}{\partial x^4}$$

where $u$ is the transverse displacement and $\kappa^2=Ek^2/\rho$, $E$ being Young's modulus, $k$ the radius of gyration of the cross-section of the bar about an axis through its centroid at right angles to the plane of bending, and $\rho$ the density.

Consider a bar of length $l$, clamped at the end $x=0$ and free at the end $x=l$, so that

$$u=\frac{\partial u}{\partial x}=0 \qquad \text{at } x=0,$$

$$\frac{\partial^2 u}{\partial x^2}=\frac{\partial^3 u}{\partial x^3}=0 \qquad \text{at } x=l.$$

Suppose that the bar is held at rest when $t<0$ with the end $x=l$ given a displacement $h$ by the application of a transverse force, which is released at $t=0$. Then the initial conditions for the motion of the bar are

$$u=u_0(x), \qquad \frac{\partial u}{\partial t}=0 \qquad \text{at } t=0,$$

where $u_0^{\text{iv}}(x)=0$, $u_0(0)=u_0'(0)=u_0''(l)=0$, $u_0(l)=h$, so that

$$u_0(x)=\tfrac{1}{2}h[3(x/l)^2-(x/l)^3].$$

The Laplace transform $u(x, p)$ satisfies

$$p^2 u - pu_0(x) = -\kappa^2 \frac{\partial^4 \bar{u}}{\partial x^4}.$$

If we put

$$\bar{u}(x, p) = u_0(x)p^{-1} + \bar{v}(x, p)$$

and set $p = 2\kappa q^2$, we have

$$\frac{\partial^4 \bar{v}}{\partial x^4} + 4q^4 \bar{v} = 0,$$

with

$$\bar{v}(0, p) = \frac{\partial \bar{v}}{\partial x}(0, p) = \frac{\partial^2 \bar{v}}{\partial x^2}(l, p) = 0,$$

$$\frac{\partial^3 \bar{v}}{\partial x^3}(l, p) = -\frac{u_0'''(l)}{p} = \frac{3h}{l^3 p}.$$

The solution of this problem (compare E2.5.5) is

$$\bar{v}(x, p)$$
$$= \frac{3h\{\cosh[q(l-x)]\cos ql \sin qx - \cos[q(l-x)]\cosh ql \sinh qx\}}{2l^3 pq^3(\cosh^2 ql + \cos^2 ql)}.$$

Since this is an even function of $q$, there is no branch point at $p = 0$, which is a simple pole with residue $-u_0(x)$.

There are also poles where $\cosh^2 ql + \cos^2 ql = 0$. The substitution $ql = \frac{1}{2}(1 + \mathrm{i})\alpha$ leads to the equation $\cosh\alpha \cos\alpha + 1 = 0$. This has an infinity of positive roots. To any root $\alpha$ there correspond roots $-\alpha$, $\pm \mathrm{i}\alpha$ and these roots give simple poles of $\bar{v}(x, p)$ at $p = \pm \mathrm{i}\alpha^2\kappa/l^2$. It can be shown that there are no other roots. The residues of $\bar{v}(x, p)$ at $p = \pm \mathrm{i}\alpha^2\kappa/l^2$ are found, after some tedious reduction, to be

$$\frac{3h\phi(x/l, \alpha)}{\alpha^4(\cosh\alpha \sin\alpha - \sinh\alpha \cos\alpha)},$$

where

$$\phi(z, \alpha) = (\sinh\alpha + \sin\alpha)(\cosh\alpha z - \cos\alpha z)$$
$$- (\cosh\alpha + \cos\alpha)(\sinh\alpha z - \sin\alpha z).$$

On applying P3.3.1 in the usual way, we get

$$u(x, t) = \sum_{n=1}^{\infty} \frac{6h\phi(x/l, \alpha_n)\cos(\alpha_n^2 \kappa t/l^2)}{\alpha_n^4(\cosh\alpha_n \sin\alpha_n - \sinh\alpha_n \cos\alpha_n)},$$

where $\alpha_n$ is the $n$th positive root of $\cos\alpha = -\operatorname{sech}\alpha$.

To justify this use of P3.3.1, we note for large $n$ that $\alpha_n \sim (n - \frac{1}{2})\pi$, so we consider the parabolae in the $p$ plane given by

$$ql = \tfrac{1}{2}N\pi(1 \pm \mathrm{i}) + \tfrac{1}{2}\xi(1 \mp \mathrm{i}), \qquad -N\pi \leqslant \xi \leqslant N\pi.$$

It can be shown that on these parabolae

$$|\cosh^2 ql + \cos^2 ql| \geqslant \cosh N\pi \cosh\xi - 1,$$

and

$$|\cosh[q(l-x)]\cos ql \sin qx - \cos[q(l-x)] \cosh ql \sinh qx| \leqslant 2\cosh N\pi \cosh\xi,$$

so that the integrand is uniformly $\mathrm{O}(|p|^{-5/2})$.

Finally, we examine the behaviour of the solution for small $t$ by expanding the transform for large $p$. We have

$$\frac{\cos[q(l-x)]\cosh ql \sinh qx - \cosh[q(l-x)]\cos ql \sin qx}{\cosh^2 ql + \cos^2 ql}$$
$$= \tfrac{1}{2}[(\mathrm{e}^{-(1+\mathrm{i})q(l-x)} + \mathrm{e}^{-(1-\mathrm{i})q(l-x)})(1 - \mathrm{e}^{-2qx} + \mathrm{e}^{-2ql} - \mathrm{e}^{-2q(l+x)})$$
$$- \mathrm{i}(\mathrm{e}^{-(1+\mathrm{i})q(l+x)} - \mathrm{e}^{-(1-\mathrm{i})q(l+x)} + \mathrm{e}^{-q(l+x)+\mathrm{i}q(l-x)}$$
$$- \mathrm{e}^{-q(l+x)-\mathrm{i}q(l-x)})(1 - \mathrm{e}^{-2q(l-x)})]$$
$$\times [1 - \mathrm{e}^{-2ql}(\mathrm{e}^{2\mathrm{i}ql} + 4 + \mathrm{e}^{-2\mathrm{i}ql}) + \dots].$$

The corresponding expansion of $v(x, t)$ will consist of a series of integrated error functions with complex arguments (E3.3.4). When $t \to 0$, all these functions will be exponentially small except those with arguments proportional to $\mathrm{e}^{\pm \mathrm{i}\pi/4}$, which can be expressed in terms of Fresnel integrals. Thus if we ignore terms leading to exponentially small functions we get

$$\bar{v}(x, p) \sim -\frac{3h(2\kappa)^{3/2}}{4l^3 p^{5/2}} \sum_{n=0}^{\infty} (-1)^n \{\exp[-(1+\mathrm{i})a_n q]$$
$$+ \exp[-(1-\mathrm{i})a_n q] - \mathrm{i}\exp[-(1+\mathrm{i})b_n q]$$
$$+ \mathrm{i}\exp[-(1-\mathrm{i})b_n q]\},$$

where

$$a_n = (2n+1)l - x, \qquad b_n = (2n+1)l + x.$$

Now

$$\mathscr{L}^{-1}\{p^{-5/2}e^{-(1+i)zq}\} = \frac{1}{6}\sqrt{\left(\frac{2}{\pi}\right)}(2t)^{3/2}\left[\left(2+\frac{iz^2}{2\kappa t}\right)\right.$$

$$\left. \times \exp\left(-\frac{iz^2}{4\kappa t}\right) - \frac{iz}{\sqrt{(2\kappa t)}}\left(3-\frac{z^2}{2\kappa t}\right)\int_{z/\sqrt{(2\kappa t)}}^{\infty} \exp(-\tfrac{1}{2}i\theta^2)d\theta\right],$$

so that as $t \to 0$

$$v(x,t) \sim -\frac{h}{\sqrt{\pi}}\left(\frac{2\kappa t}{l^2}\right)^{3/2}\sum_{n=0}^{\infty}(-1)^n$$

$$\times\left(2\cos\tfrac{1}{2}\xi_n^2 + \xi_n^2\sin\tfrac{1}{2}\xi_n^2 - \xi_n(3-\xi_n^2)\int_{\xi_n}^{\infty}\sin\tfrac{1}{2}\theta^2\,d\theta\right.$$

$$\left. -2\sin\tfrac{1}{2}\eta_n^2 + \eta_n^2\cos\tfrac{1}{2}\eta_n^2 - \eta_n(3-\eta_n^2)\int_{\eta_n}^{\infty}\cos\tfrac{1}{2}\theta^2\,d\theta\right),$$

where

$$\xi_n = \frac{a_n}{\sqrt{(2\kappa t)}}, \qquad \eta_n = \frac{b_n}{\sqrt{(2\kappa t)}}.$$

The previous two examples were broadly similar to problems on the wave equation. The following examples are on boundary layers in fluid mechanics, and so are related to the diffusion equation. The equations governing steady two-dimensional incompressible boundary layer flow are

$$u\frac{\partial u}{\partial x} + v\frac{\partial u}{\partial y} = U\frac{dU}{dx} + \nu\frac{\partial^2 u}{\partial y^2}, \tag{4.4.2}$$

$$\frac{\partial u}{\partial x} + \frac{\partial v}{\partial y} = 0. \tag{4.4.3}$$

Here $u(x, y)$ and $v(x, y)$ are the velocity components,

$$U(x) = \lim_{y\to\infty} u(x, y)$$

is the mainstream velocity, and $\nu$ is the kinematic viscosity of the fluid. Equation (4.4.3) enables a stream function $\psi(x, y)$ to be defined such that

$$u = \frac{\partial \psi}{\partial y}, \qquad v = -\frac{\partial \psi}{\partial x}.$$

**E4.4.3** In this example we shall treat the steady heat transfer from a surface through a boundary layer. The equation for the temperature $\theta(x, y)$ is

$$u\frac{\partial\theta}{\partial x}+v\frac{\partial\theta}{\partial y}=\kappa\frac{\partial^2\theta}{\partial y^2}, \tag{4.4.4}$$

where $\kappa$ is the thermometric conductivity as in Section 4.3. We seek a solution of this equation in $x>0$, $y>0$, assuming that $u$ and $v$ are known, subject to the conditions

$$\begin{aligned}&\theta=0 && \text{at } x=0,\\ &\theta=G(x) && \text{at } y=0,\\ &\theta\to 0 && \text{as } y\to\infty.\end{aligned}$$

We suppose that $y=0$ is a fixed impermeable surface, so that

$$u=v=\psi=0 \qquad \text{at } y=0.$$

When $x$ and $\psi$ are used as independent variables, equation (4.4.4) becomes

$$\frac{\partial\theta}{\partial x}=\kappa\frac{\partial}{\partial\psi}\left(u\frac{\partial\theta}{\partial\psi}\right).$$

Lighthill (1950) observed that a useful approximation near $\psi=0$, which in many cases is the most important region, is given by

$$u=\phi(x)y, \qquad \psi=\tfrac{1}{2}\phi(x)y^2,$$

where the function $\phi(x)$ is chosen from the solution of equations (4.4.2) and (4.4.3). This approximation gives

$$u=[2\phi(x)\psi]^{1/2} \qquad \text{near } \psi=0,$$

so that the approximate equation for $\theta$ is

$$\frac{\partial\theta}{\partial x}=[2\phi(x)]^{1/2}\kappa\frac{\partial}{\partial\psi}\left(\psi^{1/2}\frac{\partial\theta}{\partial\psi}\right).$$

This can be simplified by the transformation

$$\xi=\int_0^x\sqrt{[\phi(z)]}\,dz, \qquad \eta=\psi^{1/2}$$

to

$$\frac{\partial\theta}{\partial\xi}=\frac{\kappa}{2\sqrt{2}}\,\eta^{-1}\frac{\partial^2\theta}{\partial\eta^2},$$

and the boundary conditions become

$$\theta = 0 \qquad \text{at } \xi = 0,$$

$$\theta = g(\xi) \qquad \text{at } \eta = 0,$$

$$\theta \to 0 \qquad \text{as } \eta \to \infty,$$

where $g(\xi) = G(x)$.

The Laplace transform

$$\bar{\theta}(p, \eta) = \int_0^\infty \theta(\xi, \eta) \mathrm{e}^{-p\xi} \mathrm{d}\xi$$

satisfies

$$\frac{\partial^2 \bar{\theta}}{\partial \eta^2} = \frac{2\sqrt{2}}{\kappa} p\eta\bar{\theta},$$

with $\bar{\theta}(p, 0) = \bar{g}(p)$, $\bar{\theta} \to 0$ as $\eta \to \infty$.
The required solution is

$$\bar{\theta}(p, \eta) = \bar{g}(p) \frac{\mathrm{Ai}(2^{1/2}(p/\kappa)^{1/3}\eta)}{\mathrm{Ai}(0)},$$

where Ai denotes the Airy function (see E2.4.4).

The quantity of prime interest is the rate of heat transfer from the surface $y = 0$ per unit area. This is proportional to

$$-\kappa \frac{\partial \theta}{\partial y}(x, 0) = -\kappa \sqrt{[\tfrac{1}{2}\phi(x)]} \frac{\partial \theta}{\partial \eta}(\xi, 0).$$

Now

$$\frac{\partial \bar{\theta}}{\partial \eta}(p, 0) = \sqrt{2}\left(\frac{p}{\kappa}\right)^{1/3} \frac{\mathrm{Ai}'(0)}{\mathrm{Ai}(0)} \bar{g}(p) = -\sqrt{2} \frac{\Gamma(\frac{2}{3})}{\Gamma(\frac{1}{3})} \left(\frac{3p}{\kappa}\right)^{1/3} \bar{g}(p),$$

so that

$$-\frac{\partial \theta}{\partial \eta}(\xi, 0) = \frac{\sqrt{2}}{\Gamma(\frac{1}{3})} \left(\frac{3}{\kappa}\right)^{1/3} \frac{\mathrm{d}}{\mathrm{d}\xi} [\xi^{-1/3} * g(\xi)],$$

assuming that the convolution vanishes at $\xi = 0$. Consequently,

$$-\kappa \frac{\partial \theta}{\partial y}(x, 0)$$

$$= \frac{(3\kappa^2)^{1/3}}{\Gamma(\frac{1}{3})} \frac{\mathrm{d}}{\mathrm{d}x} \left[ \int_0^x G(z) \sqrt{[\phi(z)]} \left( \int_z^x \sqrt{[\phi(s)]} \mathrm{d}s \right)^{-1/3} \mathrm{d}z \right].$$

In particular, if $G(z) = G_0$ is constant we have

$$\frac{\partial\bar{\theta}}{\partial\eta}(p, 0) = -\sqrt{2}\frac{\Gamma(\frac{2}{3})}{\Gamma(\frac{1}{3})}\left(\frac{3}{\kappa}\right)^{1/3} G_0 p^{-2/3}$$

so that

$$-\frac{\partial\theta}{\partial\eta}(\xi, 0) = \frac{\sqrt{2}}{\Gamma(\frac{1}{3})}\left(\frac{3}{\kappa}\right)^{1/3} G_0 \xi^{-1/3},$$

and

$$-\kappa\frac{\partial\theta}{\partial y}(x, 0) = \frac{(3\kappa^2)^{1/3}}{\Gamma(\frac{1}{3})} G_0\sqrt{[\phi(x)]}\left(\int_0^x \sqrt{[\phi(s)]}\,\mathrm{d}s\right)^{-1/3}.$$

From this special case, the general result for arbitrary $G(x)$ can be derived as a Stieltjes integral

$$-\kappa\frac{\partial\theta}{\partial y}(x, 0) = \frac{(3\kappa^2)^{1/3}}{\Gamma(\frac{1}{3})}\sqrt{[\phi(x)]}\int_{z=0}^{z=x}\left(\int_z^x \sqrt{[\phi(s)]}\mathrm{d}s\right)^{-1/3} \mathrm{d}G(z).$$

**E4.4.4** The boundary layer equations (4.4.2) and (4.4.3) have a solution which represents the flow of a uniform mainstream over a porous surface, into which the fluid is sucked at a uniform rate. Thus $U(x) = U_0$ and $v(x, y) = -V$ are constant, and $u(x, y) = u_0(y)$. Equation (4.4.3) is satisfied automatically and equation (4.4.2) reduces to

$$-V\frac{\mathrm{d}u_0}{\mathrm{d}y} = \nu\frac{\mathrm{d}^2u_0}{\mathrm{d}y^2}.$$

The boundary conditions on $u_0(y)$ are that $u_0(0) = 0$ and $u_0(y) \to U_0$ as $y \to \infty$, so that the required solution (the *asymptotic suction profile*) is

$$u_0(y) = U_0(1 - \mathrm{e}^{-Vy/\nu}).$$

Stewartson (1957) investigated perturbations to this solution by means of the Laplace transformation. Suppose that

$$u(x, y) = u_0(y) + u_1(x, y), \qquad v(x, y) = -V + v_1(x, y),$$

where $u_1$ and $v_1$ are small relative to $u_0(y)$ and $V$, respectively. The linearized equations for $u_1$ and $v_1$ are

$$U_0(1 - \mathrm{e}^{-Vy/\nu})\frac{\partial u_1}{\partial x} - V\frac{\partial u_1}{\partial y} + U_0\frac{V}{\nu}\mathrm{e}^{-Vy/\nu}v_1 = \nu\frac{\partial^2 u_1}{\partial y^2},$$

$$\frac{\partial u_1}{\partial x} + \frac{\partial v_1}{\partial y} = 0,$$

if $U(x) = U_0$ still. In this example, we shall suppose that the perturbation is due to a change in the suction velocity in the region $x > 0$, so that the boundary conditions on the perturbation are

$$u_1 = v_1 = 0 \quad \text{at } x = 0,$$
$$u_1 = 0, \quad v_1 = -V_1(x) \quad \text{at } y = 0,$$
$$u_1 \to 0 \quad \text{as } y \to \infty.$$

It is convenient to introduce the dimensionless variables

$$\xi = \frac{V^2 x}{U_0 \nu}, \qquad \eta = \frac{V y}{\nu},$$

and to put

$$u_1 = U_0 \frac{\partial \psi}{\partial \eta}, \qquad v_1 = -V \frac{\partial \psi}{\partial \xi}, \qquad V_1(x) = V g(\xi).$$

Then the perturbation stream function $\psi(\xi, \eta)$ satisfies the equation

$$(1 - \mathrm{e}^{-\eta}) \frac{\partial^2 \psi}{\partial \xi \partial \eta} - \frac{\partial^2 \psi}{\partial \eta^2} - \mathrm{e}^{-\eta} \frac{\partial \psi}{\partial \xi} = \frac{\partial^3 \psi}{\partial \eta^3},$$

with

$$\psi(0, \eta) = 0, \qquad \frac{\partial \psi}{\partial \eta}(\xi, 0) = 0, \qquad \frac{\partial \psi}{\partial \xi}(\xi, 0) = g(\xi),$$

$$\frac{\partial \psi}{\partial \eta}(\xi, \infty) = 0.$$

We can simplify this problem by applying the operator $((\partial/\partial\eta) + 1)$ to the equation, thus obtaining

$$(1 - \mathrm{e}^{-\eta}) \frac{\partial \phi}{\partial \xi} = \frac{\partial^2 \phi}{\partial \eta^2} + \frac{\partial \phi}{\partial \eta},$$

where the new function

$$\phi(\xi, \eta) = \frac{\partial \psi}{\partial \eta} + \frac{\partial^2 \psi}{\partial \eta^2}$$

satisfies the boundary conditions

$$\phi(0, \eta) = 0, \qquad \frac{\partial \phi}{\partial \eta}(\xi, 0) = -g(\xi), \qquad \phi(\xi, \infty) = 0.$$

The Laplace transform of $\phi(\xi, \eta)$ with respect to $\xi$ satisfies

$$\frac{\partial^2 \bar{\phi}}{\partial \eta^2} + \frac{\partial \bar{\phi}}{\partial \eta} - p(1 - \mathrm{e}^{-\eta}) \bar{\phi} = 0,$$

with

$$\frac{\partial\bar{\Phi}}{\partial\eta}(p,0) = -\bar{g}(p), \qquad \bar{\Phi}(p,\infty) = 0.$$

The equation for $\bar{\Phi}(p,\eta)$ can be reduced to Bessel's equation by a suitable change of variables, and the required solution is found to be

$$\bar{\Phi}(p,\eta) = \bar{g}(p)\bar{\chi}(p,\eta),$$

where

$$\bar{\chi}(p,\eta) = \frac{2e^{-\eta/2}\,\mathrm{J}_q(2\sqrt{p}e^{-\eta/2})}{2\sqrt{p}\,\mathrm{J}'_q(2\sqrt{p}) + \mathrm{J}_q(2\sqrt{p})}$$

and $q = \sqrt{(4p+1)}$, provided $\mathscr{R}p > 0$. The function $\bar{\chi}(p,\eta)$ so defined is not singular at $p = 0$, but has a branch point at $p = -\frac{1}{4}$. It can be proved that when the $p$ plane is cut along the negative real axis from $-\infty$ to $-\frac{1}{4}$ the denominator does not vanish in the cut plane.

The asymptotic forms of Bessel functions of large order (Watson, 1944, Chapter 8; Olver, 1974, Chapter 10 Section 7 and Chapter 11 Section 10) enable us to use Jordan's lemma to express the inverse transform of $\bar{\chi}(p,\eta)$ as a loop integral round the cut. If we put $q = \pm\,\mathrm{i}r$ on the two edges of the cut, we find that

$$\chi(\xi,\eta) = \frac{e^{-\xi/4-\eta/2}}{2\pi\mathrm{i}}\int_0^\infty \left(\frac{\mathrm{I}_{-\mathrm{i}r}(\sqrt{(1+r^2)}e^{-\eta/2})}{\sqrt{(1+r^2)}\mathrm{I}'_{-\mathrm{i}r}(\sqrt{(1+r^2)})+\mathrm{I}_{-\mathrm{i}r}(\sqrt{(1+r^2)})}\right.$$
$$\left. - \frac{\mathrm{I}_{\mathrm{i}r}(\sqrt{(1+r^2)}e^{-\eta/2})}{\sqrt{(1+r^2)}\mathrm{I}'_{\mathrm{i}r}(\sqrt{(1+r^2)}) + \mathrm{I}_{\mathrm{i}r}(\sqrt{(1+r^2)})}\right)\exp(-\tfrac{1}{4}r^2\xi)r\,\mathrm{d}r,$$

and then

$$\phi(\xi,\eta) = g(\xi) * \chi(\xi,\eta).$$

In order to see how the solution behaves as $\xi \to 0$ and as $\xi \to \infty$, we must examine $\bar{\chi}(p,\eta)$ as $|p| \to \infty$ and as $p \to -\frac{1}{4}$. If $|p|$ is large and $\eta$ is not small, we have

$$\mathrm{J}_q(\sqrt{(q^2-1)}e^{-\eta/2}) \sim \frac{\exp[-q(\alpha - \tanh\alpha)]}{\sqrt{(2\pi q\tanh\alpha)}}$$

where $\operatorname{sech}\alpha = e^{-\eta/2}$, since $\sqrt{(q^2-1)} = q + \mathrm{O}(q^{-1})$. Also

$$\sqrt{(q^2-1)}\mathrm{J}'_q(\sqrt{(q^2-1)}) + \mathrm{J}_q(\sqrt{(q^2-1)}) \sim \frac{3^{1/6}\Gamma(\frac{2}{3})}{2^{1/3}\pi}q^{1/3},$$

so that in this case $\bar{\chi}(p,\eta)$ is exponentially small as $|p| \to \infty$.

However, if $\eta$ is small enough we can use the asymptotic form

$$\mathrm{J}_q(\sqrt{(q^2-1)}\mathrm{e}^{-\eta/2}) \sim \frac{1}{\pi}\sqrt{(\tfrac{1}{3}\eta)}\mathrm{K}_{1/3}(\tfrac{1}{3}q\eta^{3/2}),$$

so that

$$\bar{\chi}(p,\eta) \sim \frac{2\eta^{1/2}}{3^{1/3}\Gamma(\frac{2}{3})}p^{-1/6}\mathrm{K}_{1/3}(\tfrac{2}{3}p^{1/2}\eta^{3/2}).$$

From E1.3.9 this gives

$$\chi(\xi,\eta) \sim \frac{\xi^{-2/3}}{3^{1/3}\Gamma(\frac{2}{3})}\exp\left(-\frac{\eta^3}{9\xi}\right)$$

as $\xi \to 0$ with $\eta = \mathrm{O}(\xi^{1/3})$.

Since $q \to 0$ as $p \to -\frac{1}{4}$, we write

$$\bar{\chi}(p,\eta) = \frac{2\mathrm{e}^{-\eta/2}\mathrm{I}_q(\sqrt{(1-q^2)}\mathrm{e}^{-\eta/2})}{\sqrt{(1-q^2)}\mathrm{I}'_q(\sqrt{(1-q^2)}) + \mathrm{I}_q(\sqrt{(1-q^2)})}.$$

Now as $q \to 0$

$$\mathrm{I}_q(z) = \mathrm{I}_0(z) - q\mathrm{K}_0(z) + \mathrm{O}(q^2),$$

so that

$$\bar{\chi}(p,\eta) = 2\mathrm{e}^{-\eta/2}\left(\frac{\mathrm{I}_0(\mathrm{e}^{-\eta/2})}{I} + \frac{K\mathrm{I}_0(\mathrm{e}^{-\eta/2}) - I\mathrm{K}_0(\mathrm{e}^{-\eta/2})}{I^2}q + \mathrm{O}(q^2)\right),$$

where

$$I = \mathrm{I}'_0(1) + \mathrm{I}_0(1) \qquad K = \mathrm{K}'_0(1) + \mathrm{K}_0(1).$$

Hence when $\xi \to \infty$

$$\chi(\xi,\eta) \sim \frac{2}{\sqrt{\pi}}\xi^{-3/2}\mathrm{e}^{-\xi/4-\eta/2}\frac{I\mathrm{K}_0(\mathrm{e}^{-\eta/2}) - K\mathrm{I}_0(\mathrm{e}^{-\eta/2})}{I^2}.$$

For the special case in which $V_1(x) = Vg_0$ is constant, we have

$$\bar{\phi}(p,\eta) = g_0 p^{-1}\bar{\chi}(p,\eta).$$

Thus as $\xi \to 0$

$$\phi(\xi,\eta) \sim \frac{g_0}{3^{1/3}\Gamma(\frac{2}{3})}\int_0^{\xi}\zeta^{-2/3}\exp\left(-\frac{\eta^3}{9\zeta}\right)\mathrm{d}\zeta,$$

and as $\xi \to \infty$

$$\phi(\xi, \eta) \sim g_0\left(e^{-\eta} - \frac{\xi^{-3/2}}{2\sqrt{\pi}} e^{-\xi/4-\eta/2} \frac{I K_0(e^{-\eta/2}) - K I_0(e^{-\eta/2})}{I^2}\right),$$

since $\bar{\chi}(0, \eta) = e^{-\eta}$.

The perturbation to the velocity profile can be found from the equation

$$\frac{\partial^2 \psi}{\partial \eta^2} + \frac{\partial \psi}{\partial \eta} = \phi(\xi, \eta);$$

the most interesting quantity is the perturbation to the shear stress at $y = 0$, which is proportional to

$$\frac{\partial^2 \psi}{\partial \eta^2}(\xi, 0) = \phi(\xi, 0).$$

For the case $g(\xi) = g_0$ we have

$$\phi(\xi, 0) \sim \begin{cases} \dfrac{3^{2/3}}{\Gamma(\frac{2}{3})} g_0 \xi^{1/3} & \text{as } \xi \to 0, \\ g_0\left(1 - \dfrac{\xi^{-3/2}}{2\sqrt{\pi} I^2} e^{-\xi/4}\right) & \text{as } \xi \to \infty. \end{cases}$$

## 4.5 Additional Examples

**E4.5.1** Solve the following differential equations in the region $x > 0, t > 0$, subject to the given initial and boundary conditions:

(i) $\dfrac{\partial u}{\partial t} + \dfrac{\partial u}{\partial x} = xt, \qquad u(x, 0) = \sin x,\ u(0, t) = 0;$

(ii) $\dfrac{\partial u}{\partial t} + (x + 1)\dfrac{\partial u}{\partial x} = e^{-t}, \qquad u(x, 0) = 0,\ u(0, t) = t;$

(iii) $x\dfrac{\partial u}{\partial t} + \dfrac{\partial u}{\partial x} = x, \qquad u(x, 0) = 1,\ u(0, t) = e^t;$

(iv) $\dfrac{\partial^2 u}{\partial t^2} - \dfrac{\partial^2 u}{\partial x^2} = te^{-x}, \qquad u(x, 0) = 0,\ \dfrac{\partial u}{\partial t}(x, 0) = x, u(0, t) = 1 - e^{-t};$

(v) $\dfrac{\partial^2 u}{\partial t^2} - \dfrac{\partial^2 u}{\partial x^2} = xe^{-t}, \qquad u(x, 0) = 1,\ \dfrac{\partial u}{\partial t}(x, 0) = 0, u(0, t) = \cos t;$

(vi) $2\dfrac{\partial^2 u}{\partial t^2}+3\dfrac{\partial^2 u}{\partial x \partial t}+\dfrac{\partial^2 u}{\partial x^2}=0,$

$$u(x,0)=\frac{\partial u}{\partial t}(x,0)=u(0,t)=0,\ \frac{\partial u}{\partial x}(0,t)=1\,;$$

(vii) $\dfrac{\partial u}{\partial t}-\dfrac{\partial^2 u}{\partial x^2}=0, \qquad u(x,0)=x, \dfrac{\partial u}{\partial x}(0,t)=0\,;$

(viii) $\dfrac{\partial u}{\partial t}-\dfrac{\partial^2 u}{\partial x^2}=1, \qquad u(x,0)=0, u(0,t)=0.$

**E4.5.2** The function $u(r,t)$ satisfies the wave equation within the sphere $r=a$, that is

$$\frac{\partial^2 u}{\partial t^2}=c^2\left(\frac{\partial^2 u}{\partial r^2}+\frac{2}{r}\frac{\partial u}{\partial r}\right) \qquad (0<r<a, t>0).$$

Given that $u=0$ and $\partial u/\partial t=1$ at $t=0$, $u=0$ at $r=a$ and $u$ is finite as $r \to 0$, show that

$$\bar{u}(r,p)=p^{-2}\left(1-\frac{a}{r}\frac{\sinh(pr/c)}{\sinh(pa/c)}\right).$$

Hence obtain series expansions for $u(r,t)$.

**E4.5.3** The function $u(x,t)$ satisfies the wave equation

$$\frac{\partial^2 u}{\partial t^2}=c^2\frac{\partial^2 u}{\partial x^2} \qquad \text{for } 0<x<a, t>0.$$

Given that $u=\partial u/\partial t=0$ at $t=0, u=0$ at $x=a$, and that $u=\sin \omega t$ at $x=0$, find series expansions for $u(x,t)$. Consider separately the case in which $\omega a/c$ is an integral multiple of $\pi$.

**E4.5.4** Given that

$$\frac{\partial^2 u}{\partial t^2}=\frac{\partial^2 u}{\partial x^2}+\frac{1}{x}\frac{\partial u}{\partial x} \qquad \text{for } x>1, t>0$$

with $u=\partial u/\partial t=0$ at $t=0$, $u=1$ at $x=1$, show that

$$\bar{u}(x,p)=p^{-1}\frac{\mathrm{K}_0(xp)}{\mathrm{K}_0(p)}.$$

Deduce that for small $t$

$$u(x,t) = \left(\frac{1}{\sqrt{x}} + \frac{x-1}{8x\sqrt{x}}(t-x+1) + \mathrm{O}(t-x+1)^2\right)\mathrm{H}(t-x+1),$$

and that for large $t$

$$u(x,t) - 1 \sim v(t)\log x,$$

where

$$v(t) = \frac{1}{2\pi \mathrm{i}}\int_{-\infty}^{(0+)} \frac{\mathrm{e}^z \mathrm{d}z}{z\mathrm{K}_0(zt^{-1})}.$$

Show that $v(t) \sim -1/\log t$ as $t \to \infty$.

**E4.5.5** The temperature in a semi-infinite slab cooling by radiation from its surface satisfies the equation

$$\frac{\partial u}{\partial t} = \kappa \frac{\partial^2 u}{\partial x^2} \qquad \text{for } x > 0,\ t > 0,$$

with $u = 1$ at $t = 0$, $\partial u/\partial x = hu$ at $x = 0$. Show that

$$u(x,t) = \mathrm{erf}\left(\frac{x}{2\sqrt{(\kappa t)}}\right) + \exp(hx + h^2\kappa t)\,\mathrm{erfc}\left(\frac{x}{2\sqrt{(\kappa t)}} + h\sqrt{(\kappa t)}\right).$$

**E4.5.6** Solve

$$\frac{\partial u}{\partial t} = \frac{\partial^2 u}{\partial x^2} \qquad \text{for } x > 0,\ t > 0$$

with $u(x,0) = \mathrm{e}^{-x}$, $u(0,t) = 1$.

**E4.5.7** Given that

$$\frac{\partial u}{\partial t} = \frac{\partial^2 u}{\partial r^2} + \frac{2}{r}\frac{\partial u}{\partial r} \qquad \text{for } 0 < r < 1,\ t > 0$$

with $u = 0$ at $t = 0$, $\partial u/\partial r = 1$ at $r = 1$, $u$ finite as $r \to 0$, show that

$$u(r,t) = 3t + \tfrac{1}{2}r^2 - \tfrac{3}{10} - \frac{2}{r}\sum_{n=1}^{\infty} \frac{\sin \alpha_n r \exp(-\alpha_n^2 t)}{\alpha_n^2 \sin \alpha_n},$$

where $\alpha_1, \alpha_2, \ldots$ are the positive roots of $\tan\alpha = \alpha$.

**E4.5.8** Solve the equation

$$\frac{\partial^2 u}{\partial x \partial y} = u \qquad \text{for } x > 0,\ y > 0$$

with $u(x, 0) = e^{-x}$, $u(0, y) = 1$ in terms of a convolution by means of a Laplace transformation with respect to (i) $x$, (ii) $y$.

**E4.5.9** A transmission line of length $l$ with constant $L$, $G$, $R$ and $C$ is initially dead, that is $V = I = 0$ at $t = 0$. Given that $V(0, t) = V_0$, $I(l, t) = 0$, show that

$$V(l, t) = V_0 \operatorname{sech}[\sqrt{(RG)}l] + \frac{V_0 c^2}{l^2} e^{-\beta t} \sum_{n=0}^{\infty} \frac{(-1)^n(2n+1)\pi(\alpha_n \cos \alpha_n t + \beta \sin \alpha_n t)}{\alpha_n(\alpha_n^2 + \beta^2)},$$

where

$$\alpha_n = \sqrt{\left((n + \tfrac{1}{2})^2 \pi^2 \frac{c^2}{l^2} - k^2\right)},$$

the constants $\beta$, $k$ and $c$ are defined in E4.4.1, and $kl/c < \frac{1}{2}\pi$.

**E4.5.10** A heavy spring has mass $\rho$ per unit length when free of tension, and is attached at one end to a fixed point. When the spring is stretched its tension is $T(x, t) = k\partial y/\partial x$, where $y(x, t)$ is the extension of the portion of natural length $x$ measured from the fixed end. At $t = 0$, when the spring is hanging vertically in equilibrium, a mass $m$ is attached to the free end $x = l$ of the spring and released from rest. Show that the additional extension $z(x, t)$ satisfies the equation

$$\frac{\partial^2 z}{\partial t^2} = \frac{k}{\rho} \frac{\partial^2 z}{\partial x^2},$$

with $z = \partial z/\partial t = 0$ at $t = 0$, $z = 0$ at $x = 0$ and

$$\frac{\partial^2 z}{\partial t^2} + \frac{k}{m} \frac{\partial z}{\partial x} = g \qquad \text{at } x = l.$$

Show that the motion of the mass $m$ is given by

$$z(l, t) = \frac{mgl}{k}\left(1 - \sum_{n=1}^{\infty} \frac{2\cos(\alpha_n ct/l)}{\alpha_n^2(\alpha_n^2 \lambda^2 + \lambda + 1)}\right),$$

where $c^2 = k/\rho$, $\lambda = m/\rho l$, and $\alpha_1, \alpha_2, \ldots$ are the positive roots of $\cot \alpha = \lambda\alpha$.

**E4.5.11** A uniform chain, of length $l$ and mass $\rho$ per unit length, hangs from a fixed point and makes small transverse oscillations in a vertical plane. Show that its equation of motion is

$$\frac{\partial^2 u}{\partial t^2} = g \frac{\partial}{\partial x}\left(x \frac{\partial u}{\partial x}\right),$$

where $u(x, t)$ is the horizontal displacement at height $x$ above the free end.

The chain is released from rest with the initial displacement $u(x, 0) = \chi(x)$. Show that

$$u(x, t) = \sum_{n=1}^{\infty} a_n \mathrm{J}_0(\alpha_n \sqrt{(x/l)}) \cos[\tfrac{1}{2}\alpha_n \sqrt{(g/l)}t],$$

where $\alpha_1, \alpha_2, \ldots$ are the positive roots of $\mathrm{J}_0(\alpha) = 0$ and

$$a_n = \frac{1}{l\mathrm{J}_0'^2(\alpha_n)} \int_0^l \mathrm{J}_0(\alpha_n \sqrt{(y/l)})\chi(y)\mathrm{d}y.$$

Evaluate $a_n$ when $\chi(x) = h(l - x)/l$.

**E4.5.12** An incompressible fluid, occupying the region $y > 0$, is sucked into an infinite porous plate $y = 0$ with uniform speed $V$. At time $t = 0$, the plate is set into motion in the $x$ direction with constant speed $U$. The resultant fluid velocity is $(u(y, t), -V)$ where

$$\frac{\partial u}{\partial t} - V\frac{\partial u}{\partial y} = \nu\frac{\partial^2 u}{\partial y^2}$$

($\nu$ being the kinematic viscosity), $u(y, 0) = 0$, $u(0, t) = U$, $u(y, t) \to 0$ as $y \to \infty$. Find $u(y, t)$.

**E4.5.13** A porous circular cylinder $r = a$ is placed in an incompressible fluid which is sucked into the cylinder, so that the radial velocity is $-Va/r$. At time $t = 0$, the cylinder starts to rotate about its axis with constant angular velocity $\omega$. The resultant transverse component of velocity is then determined by

$$\frac{\partial u}{\partial t} - V\frac{a}{r}\left(\frac{\partial u}{\partial r} + \frac{u}{r}\right) = \nu\left(\frac{\partial^2 u}{\partial r^2} + \frac{1}{r}\frac{\partial u}{\partial r} - \frac{u}{r^2}\right),$$

with $u(r, 0) = 0$, $u(a, t) = a\omega$, $u(r, t)$ bounded as $r \to \infty$. Show that

$$\bar{u}(r, p) = \frac{a\omega}{p}\left(\frac{a}{r}\right)^{\lambda} \frac{\mathrm{K}_{\lambda-1}(r\sqrt{(p/\nu)})}{\mathrm{K}_{\lambda-1}(a\sqrt{(p/\nu)})},$$

where $\lambda = Va/2\nu$. Deduce that when $t \to \infty$

$$u(r, t) \to \begin{cases} a^2\omega/r & \text{if } \lambda \leqslant 1, \\ a\omega(a/r)^{2\lambda-1} & \text{if } \lambda \geqslant 1. \end{cases}$$

Find $u(r, t)$ explicitly in the cases $\lambda = \frac{1}{2}, \frac{3}{2}, \frac{5}{2}$.

**E4.5.14** $u(x, y)$ satisfies Laplace's equation

$$\frac{\partial^2 u}{\partial x^2} + \frac{\partial^2 u}{\partial y^2} = 0 \qquad \text{in } x > 0,\ 0 < y < h,$$

together with the boundary conditions $u(0, y) = 1, u(x, 0) = 0$, $u(x, h) = 0$ and $u(x, y) \to 0$ as $x \to \infty$. Show that its Laplace transform with respect to $x$ is

$$\bar{u}(p, y) = p^{-1}\left[1 - \cos py + \int_0^y f(z)\sin[p(y - z)]\,\mathrm{d}z \right.$$
$$\left. - \frac{\sin py}{\sin ph}\left(1 - \cos ph + \int_0^h f(z)\sin[p(h - z)]\,\mathrm{d}z\right)\right],$$

where

$$f(y) = \frac{\partial u}{\partial x}(0, y).$$

Use the fact that $\bar{u}(p, y)$ is regular when $\mathscr{R}p > 0$, together with Jordan's lemma, to show that

$$u(x, y) = \frac{4}{\pi}\sum_{m=0}^{\infty}\frac{1}{2m + 1}\exp\left(-(2m + 1)\pi\frac{x}{h}\right)\sin\left((2m+1)\pi\frac{y}{h}\right).$$

$$\left[u(x, y) = \frac{2}{\pi}\tan^{-1}\left\{\sin\left(\frac{\pi y}{h}\right)\operatorname{cosech}\left(\frac{\pi x}{h}\right)\right\}.\right]$$

# 5
# Further Applications

## 5.1 Integral Equations

**D5.1.1** The equations

$$\int_0^t u(s)k(t-s)\,\mathrm{d}s = g(t), \tag{5.1.1}$$

$$u(t) - \int_0^t u(s)k(t-s)\,\mathrm{d}s = g(t), \tag{5.1.2}$$

are *integral equations of convolution type* for the unknown function $u(t)$. They are examples of Volterra equations in which the *kernel* function is $k(t, s) = k(t-s)$. The functions $k(t)$ and $g(t)$ are assumed to be defined for all $t > 0$. Equations (5.1.1) and (5.1.2) are then integral equations of the *first* and *second kind*, respectively, for $u(t)$ in $t > 0$.

If $k(t)$ and $g(t)$ have Laplace transforms, the equation (5.1.2) of the second kind can be solved explicitly.

**P5.1.1** If $u(t)$ satisfies equation (5.1.2), then

$$\bar{u}(p) = \frac{\bar{g}(p)}{1-\bar{k}(p)},$$

and

$$u(t) = g(t) + \int_0^t g(s)r(t-s)\,\mathrm{d}s,$$

where the *resolvent kernel* $r(t, s) = r(t-s)$ is given by

$$r(t) = \mathscr{L}^{-1}\left\{\frac{\bar{k}(p)}{1-\bar{k}(p)}\right\}.$$

The convolution property P1.2.2 gives

$$\bar{u}(p) - \bar{u}(p)\bar{k}(p) = \bar{g}(p)$$

as the transform of equation (5.1.2), so that

$$\bar{u}(p)=\frac{\bar{g}(p)}{1-\bar{k}(p)}=\bar{g}(p)\left(1+\frac{\bar{k}(p)}{1-\bar{k}(p)}\right)=\bar{g}(p)+\bar{g}(p)\bar{r}(p).$$

From P3.2.6, $\bar{k}(p)\to 0$ uniformly as $|p|\to\infty$ in $\mathscr{R}p\geqslant A$. Hence $\bar{r}(p)$ is regular in some half plane $\mathscr{R}p\geqslant B$ and tends to zero uniformly as $|p|\to\infty$, so that its inverse transform exists.

**P5.1.2** The resolvent kernel is given by the series

$$r(t)=\sum_{n=1}^{\infty}k(t)^{*n},$$

where the convolution powers are defined in D1.5.7.

Since $|\bar{k}(p)|<K<1$ in some half plane $\mathscr{R}p\geqslant C$, the series

$$\bar{r}(p)=\sum_{n=1}^{\infty}[\bar{k}(p)]^{n}$$

converges uniformly in $\mathscr{R}p\geqslant C$.

**E5.1.1**

$$u(t)-\lambda\int_0^t u(s)\sin(t-s)\,\mathrm{d}s=g(t).$$

Here $k(t)=\lambda\sin t$, so that $\bar{k}(p)=\lambda(p^2+1)^{-1}$. Hence

$$\bar{u}(p)=\frac{\bar{g}(p)}{1-\lambda(p^2+1)^{-1}}=\bar{g}(p)\frac{p^2+1}{p^2+1-\lambda}=\bar{g}(p)+\bar{g}(p)\frac{\lambda}{p^2+1-\lambda}.$$

The resolvent kernel is therefore given by $\bar{r}(p)=\lambda(p^2+1-\lambda)^{-1}$, so that

$$u(t)=g(t)+\int_0^t g(s)r(t-s)\,\mathrm{d}s$$

where

$$r(t)=\begin{cases}\dfrac{\lambda}{\sqrt{(1-\lambda)}}\sin[\sqrt{(1-\lambda)}t] & (\lambda<1),\\ t & (\lambda=1),\\ \dfrac{\lambda}{\sqrt{(\lambda-1)}}\sinh[\sqrt{(\lambda-1)}t] & (\lambda>1).\end{cases}$$

For an equation of the first kind, the result corresponding to the first part of P5.1.1 is P5.1.3.

**P5.1.3** If

$$\int_0^t u(s)k(t-s)\,\mathrm{d}s = g(t) \tag{5.1.1}$$

then

$$\bar{u}(p) = \bar{g}(p)[\bar{k}(p)]^{-1}.$$

From P3.2.6, $[\bar{k}(p)]^{-1}$ cannot be a Laplace transform, so there is no general result analogous to the rest of P5.1.1. If the solution $u(t)$ is to exist as an ordinary function, the given function $g(t)$ must be restricted by the condition that $\bar{g}(p)[\bar{k}(p)]^{-1} \to 0$ as $|p| \to \infty$ in some half plane $\mathscr{R}p \geqslant A$. This can often be achieved by means of a condition on $g(t)$ as $t \to 0$.

**P5.1.4** If $u(t)$ satisfies equation (5.1.1), where $\bar{h}(p) = [p\bar{k}(p)]^{-1} \to 0$ uniformly as $|p| \to \infty$ in $\mathscr{R}p \geqslant A$, and $g(0)$ is finite, then

$$u(t) = \frac{\mathrm{d}}{\mathrm{d}t}\int_0^t h(s)g(t-s)\,\mathrm{d}s,$$

provided the derivative exists for all $t > 0$.

From P1.2.5, $h(t) * g(t) \to 0$ as $t \to 0$, so that if its derivative exists

$$\mathscr{L}\left\{\frac{\mathrm{d}}{\mathrm{d}t}(h(t) * g(t))\right\} = p\bar{h}(p)\bar{g}(p) = \bar{g}(p)/\bar{k}(p) = \bar{u}(p).$$

If $g'(t)$ exists for all $t > 0$, the result of P5.1.4 can also be expressed as

$$u(t) = h(t) * g'(t) + g(0)h(t).$$

**E5.1.2**

$$\int_0^t u(s)\mathrm{J}_0(t-s)\,\mathrm{d}s = g(t).$$

From E1.3.2, the transformed equation gives

$$\bar{u}(p) = (p^2+1)^{1/2}\bar{g}(p) = [p + \mathrm{O}(p^{-1})]\bar{g}(p)$$

as $|p| \to \infty$. In view of P1.2.5, we assume that $g(0) = 0$. Then

$$\bar{u}(p) = p\bar{g}(p) + [(p^2+1)^{1/2} - p]\bar{g}(p),$$

and

$$\frac{\mathrm{d}}{\mathrm{d}p}[(p^2+1)^{1/2}-p]=\frac{p}{(p^2+1)^{1/2}}-1=-\mathscr{L}\{\mathrm{J}_1(t)\}.$$

Hence from P1.3.3

$$u(t)=g'(t)+g(t)*[t^{-1}\mathrm{J}_1(t)].$$

Abel's equation is an important example of a convolution equation of the first kind.

**E5.1.3**

$$\int_0^t u(s)(t-s)^{-\nu}\,\mathrm{d}s=g(t), \qquad \text{where } 0<\mathscr{R}\nu<1.$$

Since $\mathscr{R}\nu<1$, P5.1.3 gives

$$\bar{u}(p)=\frac{p^{1-\nu}}{\Gamma(1-\nu)}\bar{g}(p)=\frac{p}{\Gamma(\nu)\Gamma(1-\nu)}\cdot\frac{\Gamma(\nu)}{p^{\nu}}\bar{g}(p).$$

Also $\mathscr{R}\nu>0$, so that $\Gamma(\nu)p^{-\nu}=\mathscr{L}\{t^{\nu-1}\}$. Thus from (3.3.2)

$$u(t)=\frac{\sin\nu\pi}{\pi}\frac{\mathrm{d}}{\mathrm{d}t}\int_0^t g(s)(t-s)^{\nu-1}\,\mathrm{d}s,$$

provided that the integral vanishes when $t\to 0$, which will be the case if $t^{\nu}g(t)\to 0$ as $t\to 0$.

If also $g(0)$ is finite and $g'(t)$ exists for $t>0$, the result can be put in the form

$$u(t)=\frac{\sin\nu\pi}{\pi}\left(g(0)t^{\nu-1}+\int_0^t g'(s)(t-s)^{\nu-1}\,\mathrm{d}s\right).$$

E5.1.3 is related to the *fractional integral* operation

$$\mathrm{I}^{\nu}u(t)=u(t)*\frac{t^{\nu-1}}{\Gamma(\nu)}=\frac{1}{\Gamma(\nu)}\int_0^t u(s)(t-s)^{\nu-1}\,\mathrm{d}s,$$

where $\mathscr{R}\nu>0$. This satisfies the functional equation

$$\mathrm{I}^{\mu}\mathrm{I}^{\nu}u(t)=\mathrm{I}^{\mu+\nu}u(t),$$

since

$$t^{\nu-1}*t^{\mu-1}=\frac{\Gamma(\mu)\Gamma(\nu)}{\Gamma(\mu+\nu)}t^{\mu+\nu-1},$$

and the solution given corresponds to the case $\mu+\nu=1$ (see E1.5.40).

Fractional integrals with other finite lower limits are easily treated by a change of origin. Another type of fractional integral is defined by

$$\mathrm{J}^{\nu}u(t) = \frac{1}{\Gamma(\nu)} \int_t^{\infty} u(s)(s-t)^{\nu-1}\,\mathrm{d}s \qquad (\mathscr{R}\nu > 0),$$

which also satisfies

$$\mathrm{J}^{\mu}\mathrm{J}^{\nu}u(t) = \mathrm{J}^{\mu+\nu}u(t).$$

Corresponding to E5.1.3 we have E5.1.4.

**E5.1.4** If

$$\int_t^{\infty} u(s)(s-t)^{-\nu}\,\mathrm{d}s = g(t),$$

where $0 < R\nu < 1$, and $t^{\lambda}g(t) \to 0$ as $t \to \infty$ where $\mathscr{R}(\lambda - \nu) > 0$, then

$$u(t) = -\frac{\sin \nu\pi}{\pi} \frac{\mathrm{d}}{\mathrm{d}t}\left(\int_t^{\infty} g(s)(s-t)^{\nu-1}\,\mathrm{d}s\right).$$

The condition on $g(t)$ ensures that the last integral converges as $s \to \infty$ and also makes it vanish when $t \to \infty$.

Equations containing derivatives can also be treated by the Laplace transformation.

**E5.1.5**

$$u'(t) - (\lambda - 1)\int_0^t u(s)\mathrm{e}^{2(t-s)}\,\mathrm{d}s = g(t).$$

The transform of the equation is

$$p\bar{u}(p) - u(0) - (\lambda - 1)\frac{\bar{u}(p)}{p-2} = \bar{g}(p),$$

so that

$$\bar{u}(p) = \frac{p-2}{p^2 - 2p - (\lambda - 1)}[\bar{g}(p) + u(0)].$$

Hence

$$u(t) = g(t) * v(t) + u(0)v(t),$$

where

$$v(t)=\begin{cases} e^t\left[\cosh(\sqrt{\lambda}t)-\dfrac{1}{\sqrt{\lambda}}\sinh(\sqrt{\lambda}t)\right] & (\lambda>0),\\ e^t(1-t) & (\lambda=0),\\ e^t\left[\cos(\sqrt{|\lambda|}t)-\dfrac{1}{\sqrt{|\lambda|}}\sin(\sqrt{|\lambda|}t)\right] & (\lambda<0). \end{cases}$$

## 5.2 Difference Equations

A *difference equation* involves the values of the unknown function at more than one argument. The simplest difference equation is

$$u(t+h)-u(t)=hg(t), \tag{5.2.1}$$

where $h$ is a positive constant. This is the analogue of the equation $du/dt=g(t)$, to which (5.2.1) reduces in the limit $h\to 0$. In order to apply the Laplace transformation to a difference equation, we need the following result.

**P5.2.1**

$$\mathscr{L}\{u(t+h)\}=e^{ph}\left(\bar{u}(p)-\int_0^h u(t)e^{-pt}\,dt\right)\qquad \text{when } h>0.$$

**P5.2.2** If $u(t)$ satisfies the difference equation (5.2.1), where $h>0$, then

$$\bar{u}(p)=\frac{\displaystyle\int_0^h u(t)e^{-pt}\,dt}{1-e^{-ph}}+\frac{h\bar{g}(p)}{e^{ph}-1}.$$

Also

$$u(t)=u_P(t)+h\sum_{n=1}^{\infty} g(t-nh)H(t-nh),$$

where

$$u_P(t)=\begin{cases} u(t) & (0\leqslant t<h),\\ u(t-h) & (t\geqslant h). \end{cases}$$

The expression for $u(t)$ comes either from P1.1.11 and P1.1.10 or from repeated application of $u(t) = u(t-h) + hg(t-h)$.

Note that the series is actually finite. The periodic function $u_P(t)$ can be chosen arbitrarily; any two solutions of (5.2.1) differ by a periodic function. The values of $u(t)$ in the range $0 \leqslant t < h$ provide the initial condition needed to define the solution $u(t)$ uniquely.

Similar results hold for difference equations with more than two values of the argument.

**E5.2.1**

$$u(t+2) - 2u(t+1) + u(t) = 1,$$

where

$$u(t) = \begin{cases} 0 & (0 < t < 1), \\ t & (1 < t < 2). \end{cases}$$

The Laplace transform of the equation is

$$e^{2p}\left(\bar{u}(p) - \int_1^2 te^{-pt}\,dt\right) - 2e^p\bar{u}(p) + \bar{u}(p) = p^{-1},$$

so that

$$\begin{aligned}(e^{2p} - 2e^p + 1)\bar{u}(p) \\ &= p^{-1} + e^{2p}[(p^{-1} + p^{-2})e^{-p} - (2p^{-1} + p^{-2})e^{-2p}] \\ &= (p^{-1} + p^{-2})(e^p - 1).\end{aligned}$$

Hence

$$\bar{u}(p) = \frac{p^{-1} + p^{-2}}{e^p - 1} = (p^{-1} + p^{-2}) \sum_{n=1}^{\infty} e^{-np},$$

which gives

$$\begin{aligned}u(t) &= \sum_{n=1}^{\infty} (1 + t - n)H(t-n), \\ &= mt - \tfrac{1}{2}m(m-1) \qquad \text{when } m < t < m+1.\end{aligned}$$

Recurrence relations connecting members of a sequence $a_n (n = 0, 1, 2, \ldots)$ can be expressed as difference equations for the function

$$\begin{aligned}u(t) &= a_n \qquad \text{when } n < t < n+1, \\ &= \sum_{n=0}^{\infty} a_n[H(t-n) - H(t-n-1)].\end{aligned}$$

**P5.2.3** If $u(t) = a_n$ when $n < t < n+1$ for $n = 0, 1, 2, \ldots$, then

$$\bar{u}(p) = \frac{1 - \mathrm{e}^{-p}}{p} \sum_{n=0}^{\infty} a_n \mathrm{e}^{-np}.$$

**E5.2.2** The Fibonacci numbers $F_n$ are defined by

$$F_{n+2} = F_{n+1} + F_n \qquad (n = 0, 1, 2, \ldots),$$
$$F_0 = 0, \qquad F_1 = 1.$$

Let $u(t) = F_n$ when $n < t < n+1$. Then

$$u(t+2) - u(t+1) - u(t) = 0 \qquad \text{for } t > 0$$

with

$$u(t) = \begin{cases} 0 & (0 < t < 1), \\ 1 & (1 < t < 2). \end{cases}$$

Thus

$$\mathrm{e}^{2p}\left(\bar{u}(p) - \int_1^2 \mathrm{e}^{-pt}\,\mathrm{d}t\right) - \mathrm{e}^{p}\bar{u}(p) - \bar{u}(p) = 0,$$

from which

$$\bar{u}(p) = \frac{1 - \mathrm{e}^{-p}}{p} \cdot \frac{\mathrm{e}^{p}}{\mathrm{e}^{2p} - \mathrm{e}^{p} - 1}.$$

To apply P5.2.3, we must express $\mathrm{e}^{p}(\mathrm{e}^{2p} - \mathrm{e}^{p} - 1)^{-1}$ as a power series in $\mathrm{e}^{-p}$. If we put $x = \mathrm{e}^{p}$,

$$\frac{x}{x^2 - x - 1} = \frac{x}{(x-\alpha)(x-\beta)} = \frac{1}{\alpha - \beta}\left(\frac{x}{x-\alpha} - \frac{x}{x-\beta}\right),$$

where $\alpha, \beta = \frac{1}{2}(1 \pm \sqrt{5})$, so that

$$\frac{x}{x^2 - x - 1} = \frac{1}{\alpha - \beta} \sum_{n=0}^{\infty} (\alpha^n - \beta^n) x^{-n}.$$

Thus

$$\bar{u}(p) = \frac{1 - \mathrm{e}^{-p}}{p} \sum_{n=0}^{\infty} F_n \mathrm{e}^{-np},$$

where

$$F_n = \frac{\alpha^n - \beta^n}{\alpha - \beta} = \frac{1}{\sqrt{5}}\left[\left(\frac{1+\sqrt{5}}{2}\right)^n - \left(\frac{1-\sqrt{5}}{2}\right)^n\right].$$

Equations involving derivatives of the unknown function may also be treated by the Laplace transformation. The next example represents a mechanical system controlled by a restoring force proportional to the displacement at an earlier time.

**E5.2.3** $u''(t) + ku'(t) + \lambda u(t-h)\mathrm{H}(t-h) = 0, (h > 0).$

Let

$$u(0) = c_0, u'(0) = c_1.$$

Then

$$p^2\bar{u} - c_0 p - c_1 + k(p\bar{u} - c_0) + \lambda \mathrm{e}^{-ph}\bar{u} = 0,$$

so that

$$\bar{u}(p) = \frac{(p+k)c_0 + c_1}{p^2 + kp + \lambda \mathrm{e}^{-ph}}.$$

The transform can be inverted as a series in two ways. If $t$ is small, it is useful to express $\bar{u}(p)$ as a power series in $\mathrm{e}^{-ph}$, thus

$$\bar{u}(p) = \sum_{n=0}^{\infty}\left(-\frac{\lambda}{k}\right)^n\left((p+k)\frac{c_0}{k} + \frac{c_1}{k}\right)\bar{u}_{n+1}(p)\mathrm{e}^{-nph},$$

where

$$\bar{u}_{n+1}(p) = \left(\frac{k}{p(p+k)}\right)^{n+1}$$

Now

$$\mathscr{L}^{-1}\left\{\frac{k}{p(p+k)}\right\} = 1 - \mathrm{e}^{-kt},$$

so that

$$u(t) = c_0 + c_1(1 - \mathrm{e}^{-kt}) + \sum_{n=1}^{\infty}\left(-\frac{\lambda}{k}\right)^n\left(\frac{c_0}{k}\frac{\mathrm{d}}{\mathrm{d}t} + c_0 + \frac{c_1}{k}\right)u_{n+1}(t-nh)\mathrm{H}(t-nh).$$

Here

$$u_{n+1}(t) = (1 - \mathrm{e}^{-kt})^{*(n+1)} = \frac{\sqrt{(\pi k)}}{n!}t^{n+1/2}\mathrm{e}^{-kt/2}\mathrm{I}_{n+1/2}(\tfrac{1}{2}kt)$$

$$= (-1/k)^n[f_n(-kt) - f_n(kt)\mathrm{e}^{-kt}],$$

where

$$f_n(x) = \sum_{m=0}^{n} \frac{(2n-m)!\,x^m}{n!\,m!\,(n-m)!}.$$

[See E1.5.17, 1.5.39, 3.5.2.]

For large values of $t$ we can use Jordan's lemma to show that $u(t)$ is the sum of the residues of $\bar{u}(p)e^{pt}$ at its poles, which are given by

$$p^2 + kp + \lambda e^{-ph} = 0. \tag{5.2.2}$$

The principle of the argument shows that, if $k$ and $\lambda$ are positive, there are two roots of (5.2.2) in the strip $|\mathscr{I}p| < \pi/h$ and one in each of the other strips of the form $(2n-1)\pi/h < \mathscr{I}p < (2n+1)\pi/h$. Thus if $q$ ranges over the roots of (5.2.2)

$$u(t) = \sum_q \frac{(q+k)c_0 + c_1}{2q + k - \lambda h e^{-qh}} e^{qt}.$$

The main interest of this problem is in whether the system is stable, that is whether $u(t) \to 0$ as $t \to \infty$ for all initial values $c_0$ and $c_1$. For stability, we require that (5.2.2) shall have no roots in $\mathscr{R}p \geqslant 0$. The principle of the argument can be used to show that the system is stable provided $\lambda > 0, k > 0$ and $\lambda h^2 < \phi(kh)$, where the

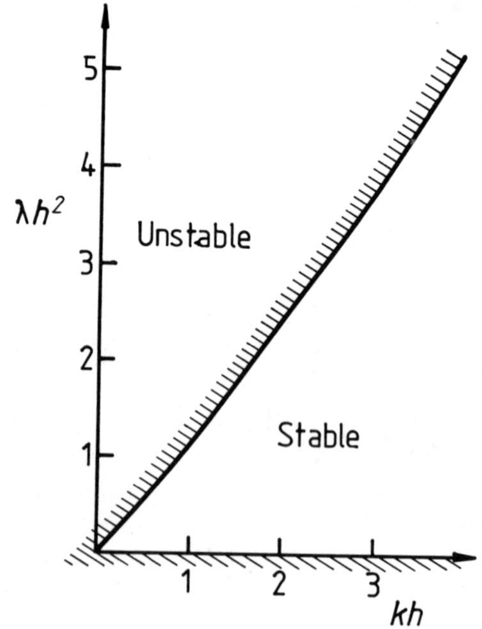


Figure 5.1

function $\phi(\xi)$ is given parametrically by $\phi(\xi)=\eta^2 \sec\eta$ when $\xi=\eta\tan\eta$ for $0<\eta<\frac{1}{2}\pi$, as shown in Figure 5.1.

We have seen that the general solution of the difference equation

$$u(t+h)-u(t)=hg(t) \tag{5.2.1}$$

contains an arbitrary periodic function. We now consider the *summation problem*, to determine a particular solution which will satisfy appropriate auxiliary conditions. Nörlund's methods for obtaining this solution are described by Milne–Thomson (1933): here we shall use the direct and inverse Laplace transformations.

Suppose first that $\mathscr{L}\{g(t)\}$ is absolutely convergent in $\mathscr{R}p \geqslant -B$, for some $B>0$. Then from P5.2.2 $\bar{u}(p)$ is regular in $\mathscr{R}p>-B$ except, in general, for poles where $e^{ph}=1$, that is at $p=2n\pi i/h$ for any integer $n$. These poles correspond to periodic fluctuations in $u(t)$, and the *principal solution* is obtained by requiring $\bar{u}(p)$ to be regular at each of these possible poles, except for $p=0$. Thus from P5.2.2

$$\int_0^h u(t)e^{-2n\pi it/h}dt = -h\bar{g}(2n\pi i/h)$$

for all integers $n\neq 0$. For $n=0$ the condition

$$\int_0^h u(t)dt=0$$

is imposed, so that the residue of $\bar{u}(p)$ at $p=0$ is

$$K=\bar{g}(0)=\int_0^\infty g(t)dt.$$

The inversion theorem P3.2.7 now gives

$$u(t)=\frac{1}{2\pi i}\int_C \frac{h\bar{g}(p)e^{pt}}{e^{ph}-1}dp,$$

where $C$ is a path in $\mathscr{R}p>-B$ from $c-i\infty$ to $c+i\infty$ $(-B<c<0)$ such that the poles $p=2n\pi i/h$ of the integrand lie to the right of $C$ for $n\neq 0$, but the pole $p=0$ is to the left of $C$.

Since $|e^{ph}|<1$ for $\mathscr{R}p<0$

$$\begin{aligned} u(t) &= K-\frac{1}{2\pi i}\int_{c-i\infty}^{c+i\infty} h\bar{g}(p)\sum_{n=0}^{\infty} e^{p(t+nh)}dp \\ &= \int_0^\infty g(t)dt - h\sum_{n=0}^{\infty} g(t+nh). \end{aligned}$$

In general, if $\mathscr{L}\{g(t)\}$ converges absolutely for $\mathscr{R}p \geqslant A$, we can define the principal solution of

$$u(t+h;\lambda) - u(t;\lambda) = h\mathrm{e}^{-\lambda t} g(t)$$

as

$$u(t;\lambda) = \frac{1}{2\pi \mathrm{i}} \int_C \frac{h\bar{g}(p+\lambda)\mathrm{e}^{pt}}{\mathrm{e}^{ph}-1} \mathrm{d}p$$

for $\mathscr{R}\lambda > A$. Since $\bar{g}(x+\mathrm{i}y) \to 0$ uniformly in $x \geqslant A$ as $y \to \pm\infty$, $u(t;\lambda)$ is a regular function of $\lambda$ and we can seek its analytic continuation to $\lambda = 0$. This can be done by modifying the path of integration provided that $\bar{g}(x+\mathrm{i}y) \to 0$ uniformly in $x \geqslant -B$ as $y \to \pm\infty$ and $\bar{g}(p)$ is regular at each of the points $2n\pi\mathrm{i}/h$ for $n \neq 0$. The analytic continuation is unambiguous for real $\lambda$ if $\bar{g}(p)$ does not have any branch point of the form $2n\pi\mathrm{i}/h + x$, where $n \neq 0$ and $x > 0$.

**D5.2.1** If $\bar{g}(x+\mathrm{i}y) \to 0$ uniformly in $x \geqslant -B$, where $B > 0$, as $y \to \pm\infty$, the *principal solution* of

$$u(t+h) - u(t) = hg(t) \tag{5.2.1}$$

is

$$u(t) = \frac{1}{2\pi \mathrm{i}} \int_C \frac{h\bar{g}(p)\mathrm{e}^{pt}}{\mathrm{e}^{ph}-1} \mathrm{d}p,$$

provided that the path $C$ can be drawn so that $\bar{g}(p)$ is regular to the right of $C$, the poles $p = 2n\pi\mathrm{i}/h$ lie to the right of $C$ except that $p = 0$ is on the left, and the $p$ plane is cut from each branch point of $\bar{g}(p)$ to $-\infty$ parallel to the real axis (see Figure 5.2).

If the condition

$$\int_0^h u(t)\mathrm{d}t = 0$$

is replaced by

$$\int_a^{a+h} u(t)\mathrm{d}t = 0,$$

the principal solution $u(t)$ of D5.2.1 is replaced by

$$u(t, a) = u(t) - \int_0^a g(t)\mathrm{d}t,$$

which is called the *sum* of $g(t)$ from $a$ to $t$.

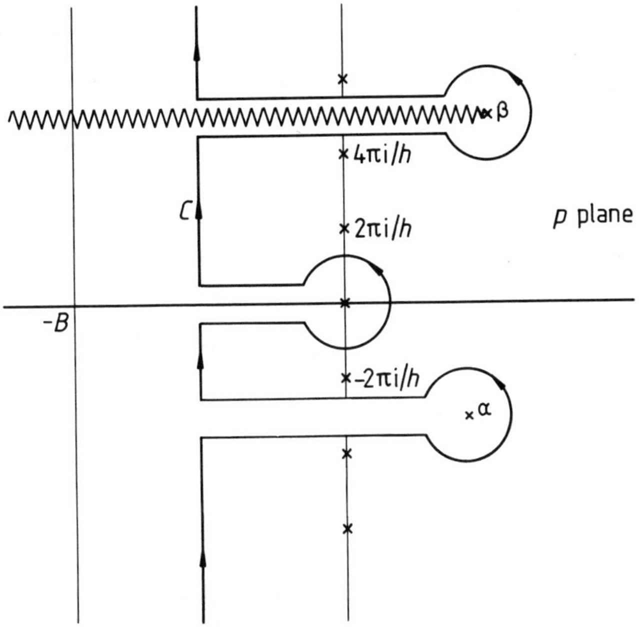


Figure 5.2

**E5.2.4** $g(t) = e^{\alpha t}$. Here $\bar{g}(p) = (p - \alpha)^{-1}$, so that the sum of $e^{\alpha t}$ from $a$ to $t$ is

$$u(t, a) = \frac{1}{2\pi i}\int_C \frac{h e^{pt}\,dp}{(p-\alpha)(e^{ph}-1)} - \int_0^a e^{\alpha t}\,dt.$$

The singularities to the left of $C$ are at $p = \alpha$ and $p = 0$. Hence

$$u(t, a) = \frac{h e^{\alpha t}}{e^{\alpha h} - 1} - \frac{1}{\alpha} - \frac{1}{\alpha}(e^{\alpha a} - 1) = \frac{h e^{\alpha t}}{e^{\alpha h} - 1} - \frac{e^{\alpha a}}{\alpha}.$$

This fails if $e^{\alpha h} = 1$, that is $\alpha = 2n\pi i/h$. The case $\alpha = 0$ gives a double pole at $p = 0$ and

$$u(t, a) = t - \tfrac{1}{2}h - a.$$

For $n \neq 0$ the contour $C$ cannot be drawn and $u(t, a)$ is undefined.

**E5.2.5** $g(t) = t^n$ $(n = 0, 1, 2, \ldots)$. Since $\bar{g}(p) = n!\,p^{-n-1}$, the sum of $t^n$ from 0 to $t$ is

$$u(t) = \frac{1}{2\pi i}\int_C \frac{n!\,h e^{pt}}{p^{n+1}(e^{ph}-1)}\,dp.$$

From D1.5.6

$$\frac{ph}{e^{ph}-1}=1-\tfrac{1}{2}ph+\sum_{m=1}^{\infty}\frac{(-1)^{m-1}B_m}{(2m)!}(ph)^{2m} \qquad (|ph|<2\pi),$$

where $B_1=\frac{1}{6}, B_2=\frac{1}{30}, B_3=\frac{1}{42}, \ldots$ are the Bernoulli numbers. Hence

$$u(t)=n!\left\{\frac{t^{n+1}}{(n+1)!}-\frac{1}{2}\frac{ht^n}{n!}+\sum_{1\leqslant m\leqslant(n+1)/2}\frac{(-1)^{m-1}B_m h^{2m}t^{n+1-2m}}{(2m)!(n+1-2m)!}\right\}.$$

When $h=1$ this is the *Bernoulli polynomial* of order $n+1$, namely

$$\phi_{n+1}(t)=\frac{t^{n+1}}{n+1}-\frac{1}{2}t^n+\sum_{1\leqslant m\leqslant(n+1)/2}\frac{(-1)^{m-1}B_m}{n+1}\binom{n+1}{2m}t^{n+1-2m}.$$

If $g(t)$ is an analytic function of $t$, regarded as a complex variable, the principal solution $u(t)$ of (5.2.1) must also be analytic. If the functions $u(t)$ and $g(t)$ are suitable, we can apply the inverse Laplace transformation to the equation. We shall therefore change the notation to

$$\bar{u}(p+\omega)-\bar{u}(p)=\omega\bar{g}(p), \tag{5.2.3}$$

where $\omega$ is a constant which may be complex.

From P3.2.6, $\bar{u}(p)\to 0$ uniformly as $p\to\infty$ in some half plane $\mathscr{R}p\geqslant A$. It follows from (5.2.3) that

$$\int_{p_0}^{p}\bar{g}(z)\mathrm{d}z\to-\frac{1}{\omega}\int_{p_0}^{p_0+\omega}\bar{u}(z)\mathrm{d}z$$

as $p\to\infty$ with both $p$ and $p+\omega$ in this half plane, and also that $p\bar{g}(p)\to 0$ as $p\to\infty$ with $|\arg(p-A)|\leqslant\frac{1}{2}\pi-\delta$.

**P5.2.4** If $\bar{u}(p)$ and $\bar{g}(p)$ are Laplace transforms that satisfy (5.2.3), then

$$u(t)=\frac{\omega}{e^{-\omega t}-1}g(t).$$

The inverse transform of (5.2.3) is $e^{-\omega t}u(t)-u(t)=\omega g(t)$.

**E5.2.6** $\psi(p)=\Gamma'(p)/\Gamma(p)$ satisfies

$$\psi(p+1)-\psi(p)=p^{-1}.$$

The condition $p\bar{g}(p) \to 0$ does not hold for this equation, but if we put

$$\bar{u}(p) = \psi(p) - \log p$$

we have

$$\bar{u}(p+1) - \bar{u}(p) = \bar{g}(p) = \frac{1}{p} - \log\left(\frac{p+1}{p}\right).$$

Now $p\bar{g}(p) \to 0$ and

$$-\bar{g}'(p) = \frac{1}{p^2} + \frac{1}{p+1} - \frac{1}{p} = \mathscr{L}\{t - 1 + \mathrm{e}^{-t}\},$$

so that (see also E1.5.16 (ii))

$$g(t) = 1 - (1 - \mathrm{e}^{-t})t^{-1}.$$

Thus from P5.2.4

$$u(t) = t^{-1} - (1 - \mathrm{e}^{-t})^{-1}.$$

Consequently

$$\psi(p) = \log p + \int_0^\infty [t^{-1} - (1 - \mathrm{e}^{-t})^{-1}]\mathrm{e}^{-pt}\,\mathrm{d}t$$

for $\mathscr{R}p > 0$. The properties of $\psi(p)$, and hence $\Gamma(p)$, may be derived from this integral representation. See also E1.5.22.

**P5.2.5** If $\bar{u}(p)$ and $\bar{g}(p)$ satisfy equation (5.2.3) where $\omega > 0$, and

$$\int_0^\infty |g(t)|^2 \mathrm{e}^{-2At}\,\mathrm{d}t$$

converges, then for $\mathscr{R}p > c > A$

$$\bar{u}(p) = \frac{1}{2\pi\mathrm{i}} \int_{c-\mathrm{i}\infty}^{c+\mathrm{i}\infty} \bar{g}(z)\left[\psi\left(\frac{p-z}{\omega}\right) - \log\left(\frac{p-z}{\omega}\right)\right]\mathrm{d}z - \int_p^\infty \bar{g}(z)\,\mathrm{d}z.$$

From P5.2.4

$$u(t) = \left(\frac{1}{t} - \frac{\omega}{1 - \mathrm{e}^{-\omega t}}\right)g(t) - \frac{g(t)}{t},$$

and the result follows by combining E1.5.22, P5.4.10 and P1.3.3.

**P5.2.6** With the conditions of P5.2.5

$$\bar{u}(p) = \frac{1}{2\pi i}\int_{k-i\infty}^{k+i\infty} G(p-\omega\zeta)\left(\frac{\pi}{\sin \pi\zeta}\right)^2 d\zeta,$$

where

$$G(p) = \int_{\infty}^{p} \bar{g}(z)dz, \qquad 0 < k < 1 \text{ and } \mathscr{R}p > A + k\omega.$$

Integration by parts from P5.2.5 gives

$$\bar{u}(p) = \frac{1}{2\pi i\omega}\int_{c-i\infty}^{c+i\infty} G(z)\psi'\left(\frac{p-z}{\omega}\right)dz,$$

since we can complete the contour by a semicircle on the right. Also $\psi'(z-p)/(\omega+1)$ is regular in $\mathscr{R}z > \mathscr{R}p - \omega$ and is $O(z^{-1})$ at infinity, so that

$$\int_{c-i\infty}^{c+i\infty} G(z)\psi'\left(\frac{z-p}{\omega}+1\right)dz = 0$$

provided that $c > \mathscr{R}p - \omega$. Combining these results with the aid of

$$\psi'(\zeta) + \psi'(1-\zeta) = \pi^2 \operatorname{cosec}^2 \pi\zeta,$$

which is a consequence of equation (3.3.2), we obtain

$$\bar{u}(p) = \frac{1}{2\pi i\omega}\int_{c-i\infty}^{c+i\infty} G(z)\pi^2 \operatorname{cosec}^2\left(\frac{\pi}{\omega}(p-z)\right)dz,$$

from which the given result follows on putting $z = p - \omega\zeta$.

This result can now be generalized to functions $\bar{g}(p)$ and $\bar{u}(p)$ that are not Laplace transforms. If $G(p)$ is replaced by $G(p) + K$, where $K$ is a constant, the effect is to change $\bar{u}(p)$ into $\bar{u}(p) + K$. Thus we have D5.2.2.

**D5.2.2** If $\bar{g}(p)$ is regular for all finite $p$, $\bar{g}(p) = O(e^{b|p|})$ for large $|p|$, and

$$G(p, a) = \int_{a}^{p} \bar{g}(z)dz,$$

the *sum* of $\bar{g}(p)$ from $a$ to $p$ is given by

$$\bar{u}(p, a) = \frac{1}{2\pi i}\int_{k-i\infty}^{k+i\infty} G(p-\omega\zeta, a)\left(\frac{\pi}{\sin \pi\zeta}\right)^2 d\zeta,$$

where $0 < k < 1$ and $\omega$, which may be complex, is such that

$$|\omega| < 2\pi/b.$$

**E5.2.7** $\bar{g}(p) = e^{\alpha p}$ gives $G(p, a) = \alpha^{-1}(e^{\alpha p} - e^{\alpha a})$ so that

$$\bar{u}(p, a) = \frac{1}{2\pi i}\int_{k-i\infty}^{k+i\infty} \alpha^{-1}(e^{\alpha(p-\omega\zeta)} - e^{\alpha a})\left(\frac{\pi}{\sin \pi\zeta}\right)^2 d\zeta.$$

The integral converges provided that $|\mathscr{I}(\alpha\omega)| < 2\pi$. The constant term $(-\alpha^{-1}e^{\alpha a})$ of $G(p, a)$ contributes the same amount to $\bar{u}(p, a)$. To evaluate the remainder, put $\zeta = Z + 1$ so that

$$\bar{u}(p, a) + \alpha^{-1}e^{\alpha a}$$

$$= \frac{1}{2\pi i}\int_{k-1-i\infty}^{k-1+i\infty} \alpha^{-1}e^{\alpha(p-\omega Z-\omega)}\left(\frac{\pi}{\sin \pi Z}\right)^2 dZ$$

$$= \frac{1}{2\pi i}\int_{k-i\infty}^{k+i\infty} \alpha^{-1}e^{\alpha(p-\omega Z-\omega)}\left(\frac{\pi}{\sin \pi Z}\right)^2 dZ + \omega e^{\alpha(p-\omega)}$$

on moving the contour across the pole at $Z = 0$. Thus

$$(1 - e^{-\alpha\omega})[\bar{u}(p, a) + \alpha^{-1}e^{\alpha a}] = \omega e^{\alpha(p-\omega)},$$

and

$$\bar{u}(p, a) = \frac{\omega e^{\alpha p}}{e^{\alpha\omega} - 1} - \frac{e^{\alpha a}}{\alpha}.$$

Note that this result agrees with E5.2.4. As a function of $\omega$, $\bar{u}(p, a)$ is regular except for simple poles at $\omega = 2n\pi i/\alpha$ for $n \neq 0$, and $\bar{u}(p, a)$ reduces to $G(p, a)$ as $\omega \to 0$. $\bar{u}(p, a)$ is also regular as a function of $\alpha$, except for simple poles at $\alpha = 2n\pi i/\omega\,(n \neq 0)$, and the coefficient of $\alpha^n$ in the power series expansion about $\alpha = 0$ gives the result corresponding to E5.2.5.

## 5.3 Fourier Series and Integrals

The Laplace transform of a periodic function was obtained in P1.1.11, which may be stated as follows.

**P5.3.1** If $v(t) = 0$ for $t > T$ and

$$u(t) = \begin{cases} v(t) & (0 < t < T), \\ u(t - T) & (t > T), \end{cases}$$

then

$$\bar{u}(p) = (1 - e^{-pT})^{-1}\bar{v}(p).$$

The function $\bar{v}(p)$ is regular for all finite $p$, so that the only singularities of $\bar{u}(p)$ are simple poles where $e^{-pT} = 1$. The residue of $\bar{u}(p)$ at $p = 2n\pi i/T$ is $T^{-1}\bar{v}(2n\pi i/T)$.

**P5.3.2** If $u(t) = u(t - T)$ for *all* $t$ and $u_1(t) = u(T - t)$, then $\bar{u}_1(p) = -\bar{u}(-p)$.

Let $v(t) = u(t)H(T - t)$, as in P5.3.1. Then

$$(1 - e^{-pT})\bar{u}_1(p) = \int_0^T v(T - t)e^{-pt}\,dt = \int_0^T v(s)e^{-p(T-s)}\,ds$$
$$= e^{-pT}\bar{v}(-p),$$

so that

$$\bar{u}_1(p) = \frac{\bar{v}(-p)}{e^{pT} - 1} = -\bar{u}(-p).$$

From the Laplace transform $\bar{u}(p)$, we obtain the Fourier series of $u(t)$ as follows.

**P5.3.3** Let $u(t)$ be integrable in $0 < t < T$ except in the neighbourhoods of a finite number of points, and let

$$\int_0^T |u(t)|\,dt$$

exist, if necessary as an improper integral. Define $u(t)$ outside the range $0 < t < T$ by $u(t) = u(t - T)$. If $u(\tau \pm 0)$ exist such that as $s \to 0+$

$$|u(\tau \pm s) - u(\tau \pm 0)| = O(s^a), \qquad \text{where } a > 0,$$

then

$$\lim_{N\to\infty} \sum_{n=-N}^{N} c_n e^{2n\pi i\tau/T} = \tfrac{1}{2}[u(\tau + 0) + u(\tau - 0)],$$

where

$$c_n = T^{-1}\int_0^T u(t)e^{-2n\pi it/T}\,dt.$$

Consider

$$I(Y) = \frac{1}{2\pi i}\int_C \bar{u}(p)e^{p\tau}\,dp,$$

where $C$ is the rectangle with corners $\pm c \pm \mathrm{i}Y$, and $Y = (2N+1)\pi/T$. From P5.3.1

$$I(Y) = \sum_{n=-N}^{N} T^{-1}\bar{v}\left(\frac{2n\pi\mathrm{i}}{T}\right)\mathrm{e}^{2n\pi\mathrm{i}\tau/T} = \sum_{n=-N}^{N} c_n \mathrm{e}^{2n\pi\mathrm{i}\tau/T},$$

where $v(t) = u(t)\mathrm{H}(T-t)$.

From P3.2.5, $\bar{v}(x \pm \mathrm{i}y) \to 0$ uniformly as $y \to \infty$ for $x \geqslant -c$. On $y = \pm Y$, $\bar{u}(x+\mathrm{i}y) = \frac{1}{2}\bar{v}(x+\mathrm{i}y)$ so that the integrals along the top and bottom of $C$ tend to zero as $N \to \infty$.

The inversion theorem P3.2.7 shows that as $Y \to \infty$

$$\frac{1}{2\pi\mathrm{i}}\int_{c-\mathrm{i}Y}^{c+\mathrm{i}Y} \bar{u}(p)\mathrm{e}^{p\tau}\,\mathrm{d}p \to \tfrac{1}{2}[u(\tau+0) + u(\tau-0)]$$

if $\tau > 0$. Also

$$\frac{1}{2\pi\mathrm{i}}\int_{-c+\mathrm{i}Y}^{-c-\mathrm{i}Y} \bar{u}(p)\mathrm{e}^{p\tau}\,\mathrm{d}p = -\frac{1}{2\pi\mathrm{i}}\int_{c-\mathrm{i}Y}^{c+\mathrm{i}Y} \bar{u}(-z)\mathrm{e}^{-z\tau}\,\mathrm{d}z$$

$$= \frac{1}{2\pi\mathrm{i}}\int_{c-\mathrm{i}Y}^{c+\mathrm{i}Y} \bar{u}_1(z)\mathrm{e}^{-z\tau}\,\mathrm{d}z,$$

where $u_1(t) = u(T-t)$, from P5.3.2. Then P3.2.7 shows that if $\tau > 0$ this integral tends to zero as $Y \to \infty$. The result extends to all $\tau$ by periodicity.

Note that the series does not necessarily converge separately as $n \to \pm\infty$ (compare P3.2.7). From the complex-exponential form of the Fourier series, the trigonometrical form is obtained by writing $c_n = \frac{1}{2}(a_n - \mathrm{i}b_n)$, $c_{-n} = \frac{1}{2}(a_n + \mathrm{i}b_n)$.

**P5.3.4** With the hypotheses of P5.3.3

$$\tfrac{1}{2}a_0 + \sum_{n=1}^{\infty}\left[a_n\cos\left(\frac{2n\pi\tau}{T}\right) + b_n\sin\left(\frac{2n\pi\tau}{T}\right)\right]$$

$$= \tfrac{1}{2}[u(\tau+0) + u(\tau-0)],$$

where

$$a_n = 2T^{-1}\int_0^T u(t)\cos\left(\frac{2n\pi t}{T}\right)\mathrm{d}t,$$

$$b_n = 2T^{-1}\int_0^T u(t)\sin\left(\frac{2n\pi t}{T}\right)\mathrm{d}t.$$

As in the case of the inverse Laplace transform, the results of P5.3.3 and P5.3.4 hold for all $\tau$ if $u(t)$ is piecewise differentiable.

**E5.3.1** For the square-wave function (E1.1.12)

$$u(t) = \begin{cases} 1 & (nT < t < (n+\tfrac{1}{2})T), \\ 1 & ((n+\tfrac{1}{2})T < t < (n+1)T), \end{cases}$$

we have

$$v(t) = \mathrm{H}(t) - 2\mathrm{H}(t - \tfrac{1}{2}T) + \mathrm{H}(t - T),$$

so that

$$\bar{v}(p) = (1 - 2\mathrm{e}^{-pT/2} + \mathrm{e}^{-pT})p^{-1} = (1 - \mathrm{e}^{-pT/2})^2 p^{-1}.$$

The Fourier coefficients of $u(t)$ are therefore given by

$$c_n = T^{-1}(1 - \mathrm{e}^{-n\pi \mathrm{i}})^2 \frac{T}{2n\pi \mathrm{i}},$$

so that

$$c_{2m} = 0, \qquad c_{2m+1} = \frac{2}{(2m+1)\pi \mathrm{i}},$$

and hence

$$a_n = 0, \qquad b_{2m} = 0 \qquad \text{and} \qquad b_{2m+1} = \frac{4}{(2m+1)\pi}.$$

Some of the solutions of partial differential equations given in Chapter 4 are in the form of Fourier series.

The inversion theorem for the Laplace transform is often deduced from that for the Fourier transform, but this inference can be reversed.

**D5.3.1** Let $u(t)$ be integrable in $-X \leqslant t \leqslant Y$, except in the neighbourhoods of a finite number of points, for all finite $X$, $Y$ and let

$$\int_{-X}^{Y} |u(t)| \, \mathrm{d}t$$

exist for all finite $X$, $Y$ and have a finite limit as $X$ and $Y$ tend to infinity. The Fourier transform of $u(t)$ is defined for real $k$ as

$$U(k) = \int_{-\infty}^{\infty} u(t)\mathrm{e}^{-\mathrm{i}kt} \, \mathrm{d}t.$$

**P5.3.5** With $u(t)$ as in D5.3.1, let $u_1(t) = u(-t)$. Then

$$U(k) = \bar{u}(\mathrm{i}k) + \bar{u}_1(-\mathrm{i}k).$$

From D5.3.1 the Laplace transforms $\bar{u}(p)$ and $\bar{u}_1(p)$ converge absolutely when $\mathscr{R}p = 0$.

P5.3.5 represents $U(k)$ as the sum of two functions, of which $\bar{u}(\mathrm{i}k)$ is regular in $\mathscr{I}k < 0$ and $\bar{u}_1(-\mathrm{i}k)$ is regular in $\mathscr{I}k > 0$. In general, these functions may have singularities when $k$ is real, so that $U(k)$ will not be analytic on the whole real axis. If, however, $u(t) = \mathrm{O}(\mathrm{e}^{-A|t|})$ for some $A > 0$ as $t \to \pm\infty$, $\bar{u}(\mathrm{i}k)$ is regular in $\mathscr{I}k < A$ and $\bar{u}_1(-\mathrm{i}k)$ in $\mathscr{I}k > -A$, so that $U(k)$ is then regular in the strip $|\mathscr{I}k| < A$.

**P5.3.6** If $u(t)$ satisfies the conditions of D5.3.1 and $u(\tau \pm 0)$ exist such that $|u(\tau \pm s) - u(\tau \pm 0)| = \mathrm{O}(s^a)$ as $s \to 0+$, where $a > 0$, then

$$\lim_{X \to \infty} \frac{1}{2\pi} \int_{-X}^{X} U(k)\mathrm{e}^{\mathrm{i}k\tau}\,\mathrm{d}k = \tfrac{1}{2}[u(\tau + 0) + u(\tau - 0)].$$

Let $u_1(t) = u(-t)$ and

$$I(X) = \frac{1}{2\pi} \int_{-X}^{X} \bar{u}(\mathrm{i}k)\mathrm{e}^{\mathrm{i}k\tau}\,\mathrm{d}k, \qquad I_1(X) = \frac{1}{2\pi} \int_{-X}^{X} \bar{u}_1(\mathrm{i}k)\mathrm{e}^{-\mathrm{i}k\tau}\,\mathrm{d}k.$$

Then from P3.2.7, as $X \to \infty$

$$I(X) \to \tfrac{1}{2}[u(\tau + 0) + u(\tau - 0)], \qquad \text{if } \tau > 0;$$
$$I(X) \to 0,\ I_1(X) \to \tfrac{1}{2}[u_1(-\tau + 0) + u_1(-\tau - 0)] \text{ if } \tau < 0;$$
$$I(X) \to \tfrac{1}{2}u(0+),\ I_1(X) \to \tfrac{1}{2}u_1(0+) \quad \text{if } \tau = 0.$$

The result follows from P5.3.5.

As with the Fourier series, P5.3.6 applies to all $\tau$ if $u(t)$ is piecewise differentiable.

**E5.3.2** $u(t) = \mathrm{e}^{-|t|}$ gives $\bar{u}(p) = \bar{u}_1(p) = (p + 1)^{-1}$. $\bar{u}(\mathrm{i}k) = (\mathrm{i}k + 1)^{-1}$ is regular in $\mathscr{I}k < 1$, $\bar{u}_1(-\mathrm{i}k) = (-\mathrm{i}k + 1)^{-1}$ is regular in $\mathscr{I}k > -1$, and $U(k) = 2(k^2 + 1)^{-1}$ is regular in $|\mathscr{I}k| < 1$.

**E5.3.3** $u(t) = (t^2 + 1)^{-1}$ gives (see E1.3.8)

$$\bar{u}(p) = \bar{u}_1(p) = \int_0^\infty \frac{p}{p^2 + r^2}\mathrm{e}^{-r}\,\mathrm{d}r = \int_p^\infty \frac{\sin(z - p)}{z}\,\mathrm{d}z.$$

This is regular except at $p = 0$, which is a branch point. Near $p = 0$,

$$\bar{u}(p) = \tfrac{1}{2}\pi + p(\log p + \gamma - 1) + O(p^2).$$

When $p = ik$ for real $k \neq 0$

$$\bar{u}(ik) = ik\,P\int_0^\infty \frac{e^{-r}}{r^2 - k^2}\,dr + \tfrac{1}{2}\pi e^{-|k|},$$

where P denotes a principal value at $r = k$, so that

$$U(k) = \bar{u}(ik) + \bar{u}_1(-ik) = \pi e^{-|k|}.$$

There is no strip of regularity in this case as the functions $\bar{u}(ik)$ and $\bar{u}_1(-ik)$ are both singular at $k = 0$.

## 5.4 The Parseval Formulae

Laplace transforms have properties analogous to Parseval's theorem on Fourier series. The heuristic argument in the simplest case is as follows. Suppose that $u(t)$ is real for $t > 0$. Then $\bar{u}(x - iy)$ is the complex conjugate of $\bar{u}(x + iy)$ and

$$\begin{aligned}
\int_{-\infty}^{\infty} |\bar{u}(x+iy)|^2\,dy &= \int_{-\infty}^{\infty} \bar{u}(x+iy)\bar{u}(x-iy)\,dy \\
&= \int_{-\infty}^{\infty} \bar{u}(x+iy)\int_0^\infty u(t)e^{-(x-iy)t}\,dt\,dy \\
&= \int_{-\infty}^{\infty} u(t)\int_0^\infty \bar{u}(x+iy)e^{(x+iy)t}\,dy\,e^{-2xt}\,dt \\
&= \int_0^\infty u(t)\cdot 2\pi u(t)\cdot e^{-2xt}\,dt \\
&= 2\pi\int_0^\infty u^2(t)e^{-2xt}\,dt.
\end{aligned}$$

Clearly we must assume that the last integral converges, but it would be difficult to justify the argument without making restrictive assumptions also about the transform $\bar{u}(p)$, and in order to prove the result we shall proceed indirectly.

**P5.4.1** If

$$\int_0^\infty |u(t)|^2 e^{-2At}\,dt$$

converges for some real $A$ then

$$\bar{u}(p) = \int_0^\infty u(t)e^{-pt}\,dt$$

converges absolutely when $\mathscr{R}p > A$.

For

$$\left(\int_0^T |u(t)e^{-pt}|\,dt\right)^2 = \left(\int_0^T |u(t)|e^{-At}\cdot\exp[-(\mathscr{R}p - A)t]\,dt\right)^2$$

$$\leqslant \int_0^T |u(t)|^2 e^{-2At}\,dt \cdot \int_0^T \exp[-2(\mathscr{R}p - A)t]\,dt$$

by Schwarz's inequality.

**E5.4.1** If $u(t) = (t+1)^{-1}$,

$$\int_0^\infty |u(t)|^2 e^{-2xt}\,dt$$

converges for $x \geqslant 0$, and

$$\int_0^\infty |u(t)e^{-pt}|\,dt$$

converges for $\mathscr{R}p > 0$.

**P5.4.2** Let $u(t)$ be real,

$$\int_0^\infty u^2(t)e^{-2At}\,dt$$

converge, and $x > A$. Then

$$J(\delta) = \int_{-\infty}^\infty \exp(-\delta^2 y^2)|\bar{u}(x + iy)|^2\,dy$$

converges for $\delta > 0$ and

$$J(\delta) = 4\sqrt{\pi}\int_0^\infty \exp(-\xi^2)\Phi(2\delta\xi)\,d\xi,$$

where

$$\Phi(\theta) = \int_0^\infty u(t)e^{-xt}\cdot u(t+\theta)e^{-x(t+\theta)}\,dt,$$

Also

$$|\Phi(\theta)| \leqslant \Phi(0) \qquad \text{and} \qquad J(\delta) \leqslant 2\pi\Phi(0).$$

Since $x > A, \bar{u}(x+\mathrm{i}y)$ is an absolutely convergent integral and tends to zero as $y \to \pm\infty$. Hence $|\bar{u}(x+\mathrm{i}y)|^2$ is bounded and so $J(\delta)$ converges. Because of the absolute convergence, we can change the orders of integration so that

$$\begin{aligned}
J(\delta) &= \int_{-\infty}^{\infty} \exp(-\delta^2 y^2) \int_0^{\infty} u(t)\mathrm{e}^{-(x+\mathrm{i}y)t}\,\mathrm{d}t \int_0^{\infty} u(s)\mathrm{e}^{-(x-\mathrm{i}y)s}\,\mathrm{d}s\,\mathrm{d}y \\
&= \int_0^{\infty}\int_0^{\infty} \mathrm{e}^{-x(s+t)}u(s)u(t)\int_{-\infty}^{\infty} \exp[-\delta^2y^2+\mathrm{i}(s-t)y]\,\mathrm{d}y\,\mathrm{d}s\,\mathrm{d}t \\
&= \int_0^{\infty}\int_0^{\infty} \mathrm{e}^{-x(s+t)}u(s)u(t)\frac{\sqrt{\pi}}{\delta}\exp\left[-\left(\frac{s-t}{2\delta}\right)^2\right]\mathrm{d}s\,\mathrm{d}t \\
&= 2\sqrt{\pi}\int_{-\infty}^{\infty} \exp(-\xi^2)\int_{\delta|\xi|}^{\infty} \mathrm{e}^{-2x\tau}u(\tau-\delta\xi)u(\tau+\delta\xi)\,\mathrm{d}\tau\,\mathrm{d}\xi,
\end{aligned}$$

where $s-t = 2\delta\xi$, $s+t = 2\tau$. The inner integral is an even function of $\xi$ and is equal to $\Phi(2\delta|\xi|)$. Thus

$$\mathrm{J}(\delta) = 4\sqrt{\pi}\int_0^{\infty} \exp(-\xi^2)\Phi(2\delta\xi)\,\mathrm{d}\xi \qquad \text{for } \delta > 0.$$

From Schwarz's inequality,

$$\begin{aligned}
\Phi^2(\theta) &\leqslant \int_0^{\infty} u^2(t)\mathrm{e}^{-2xt}\,\mathrm{d}t \int_0^{\infty} u^2(s+\theta)\mathrm{e}^{-2x(s+\theta)}\,\mathrm{d}s \\
&\leqslant \left(\int_0^{\infty} u^2(t)\mathrm{e}^{-2xt}\,\mathrm{d}t\right)^2 = \Phi^2(0).
\end{aligned}$$

Since $\Phi(0) \geqslant 0$ this gives $|\Phi(\theta)| \leqslant \Phi(0)$ and so

$$J(\delta) \leqslant 4\sqrt{\pi}\int_0^{\infty} \exp(-\xi^2)\Phi(0)\,\mathrm{d}\xi = 2\pi\Phi(0).$$

**P5.4.3** $\Phi(\theta)$ is continuous. In particular $\Phi(\theta) \to \Phi(0)$ as $\theta \to 0+$.

Put $v(t) = u(t)\mathrm{e}^{-xt}$, and suppose that $0 \leqslant \theta < k$, $0 < \theta + \eta < k$. We have to show that as $\eta \to 0$

$$\Phi(\theta+\eta) - \Phi(\theta) = \int_0^{\infty} v(t)\,[v(t+\theta+\eta) - v(t+\theta)]\,\mathrm{d}t \to 0.$$

First choose $T$ such that

$$\int_T^{\infty} v^2(t)\,\mathrm{d}t < \varepsilon.$$

Then if $\lambda \geqslant 0$

$$\left(\int_T^\infty v(t)v(t+\lambda)\,\mathrm{d}t\right)^2 \leqslant \int_T^\infty v^2(t)\,\mathrm{d}t \int_T^\infty v^2(s+\lambda)\,\mathrm{d}s < \varepsilon^2,$$

so that

$$\left|\int_T^\infty v(t)v(t+\lambda)\,\mathrm{d}t\right| < \varepsilon \qquad \text{for all } \lambda \geqslant 0.$$

Hence

$$\left|\int_T^\infty v(t)[v(t+\theta+\eta)-v(t+\theta)]\,\mathrm{d}t\right| \leqslant 2\varepsilon.$$

If $v(t)$ is not properly integrable in the neighbourhoods of certain points in $0 \leqslant t \leqslant T$, we can enclose these points in a finite set of intervals $I$ such that

$$\int_I v^2(t)\,\mathrm{d}t < \varepsilon^2\left(\int_0^\infty v^2(s)\,\mathrm{d}s\right)^{-1}$$

Then

$$\left(\int_I v(t)v(t+\lambda)\,\mathrm{d}t\right)^2 \leqslant \int_I v^2(t)\,\mathrm{d}t \int_I v^2(s+\lambda)\,\mathrm{d}s$$

$$\leqslant \int_I v^2(t)\,\mathrm{d}t \int_0^\infty v^2(s)\,\mathrm{d}s < \varepsilon^2.$$

Hence as before

$$\left|\int_I v(t)[v(t+\theta+\eta)-v(t+\theta)]\,\mathrm{d}t\right| < 2\varepsilon.$$

In $J$, the remainder of $0 \leqslant t \leqslant T$ after removal of the intervals $I$, $v(t)$ is bounded, say $|v(t)| \leqslant M$. Then

$$\left|\int_J v(t)[v(t+\theta+\eta)-v(t+\theta)]\,\mathrm{d}t\right|$$

$$\leqslant \int_J M|v(t+\theta+\eta)-v(t+\theta)|\,\mathrm{d}t$$

$$\leqslant M\int_0^T |v(t+\theta+\eta)-v(t+\theta)|\,\mathrm{d}t.$$

From P1.2.3, since $v(t)$ is absolutely integrable in $0 \leqslant t \leqslant T+k$, we can find $\eta_0(\varepsilon)$ such that

$$\int_0^T |v(t+\theta+\eta) - v(t+\theta)|\,dt < \varepsilon/M \qquad \text{when } |\eta| \leqslant \eta_0(\varepsilon).$$

Hence the result.

**P5.4.4** As $\delta \to 0+$,

$$\int_0^\infty \exp(-\xi^2)\Phi(2\delta\xi)\,d\xi \to \tfrac{1}{2}\sqrt{(\pi)}\Phi(0).$$

Since (P5.4.2) $|\Phi(\theta)| \leqslant \Phi(0)$

$$0 \leqslant \int_X^\infty \exp(-\xi^2)[\Phi(0) - \Phi(2\delta\xi)]\,d\xi$$

$$\leqslant 2\Phi(0)\int_X^\infty \exp(-\xi^2)\,d\xi < \varepsilon,$$

if we choose $X$ sufficiently great. Also (P5.4.3)

$$0 \leqslant \Phi(0) - \Phi(\theta) < \varepsilon \qquad \text{when } 0 \leqslant \theta \leqslant \theta_0(\varepsilon),$$

so that if $0 \leqslant 2\delta X \leqslant \theta_0(\varepsilon)$

$$0 \leqslant \int_0^X \exp(-\xi^2)[\Phi(0) - \Phi(2\delta\xi)]\,d\xi$$

$$< \varepsilon\int_0^X \exp(-\xi^2)\,d\xi < \tfrac{1}{2}\sqrt{(\pi)}\varepsilon.$$

Hence when $\delta \to 0+$

$$\int_0^\infty \exp(-\xi^2)\Phi(2\delta\xi)\,d\xi \to \int_0^\infty \exp(-\xi^2)\Phi(0)\,d\xi = \tfrac{1}{2}\sqrt{\pi}\,\Phi(0).$$

**P5.4.5**

$$\int_{-\infty}^\infty |\bar{u}(x+iy)|^2\,dy$$

converges to

$$2\pi\Phi(0) = 2\pi\int_0^\infty u^2(t)e^{-2xt}\,dt.$$

For any positive $\delta$ and $Y$

$$\int_{-\infty}^{\infty}|\bar{u}(x+\mathrm{i}y)|^2\,\mathrm{d}y \geqslant J(\delta) \geqslant \int_{-Y}^{Y}|\bar{u}(x+\mathrm{i}y)|^2\exp(-\delta^2y^2)\,\mathrm{d}y$$

$$\geqslant \exp(-\delta^2Y^2)\int_{-Y}^{Y}|\bar{u}(x+\mathrm{i}y)|^2\,\mathrm{d}y.$$

From P5.4.2 and P5.4.4, $J(\delta)\to 2\pi\Phi(0)$ as $\delta\to 0+$, so that

$$\int_{-\infty}^{\infty}|\bar{u}(x+\mathrm{i}y)|^2\,\mathrm{d}y \geqslant 2\pi\Phi(0) \geqslant \int_{-Y}^{Y}|\bar{u}(x+\mathrm{i}y)|^2\,\mathrm{d}y$$

for all $Y>0$. Hence the result.

**E5.4.2** $u(t)=t^{\nu}$ gives

$$\int_0^{\infty}u^2(t)\mathrm{e}^{-2xt}\,\mathrm{d}t=\Gamma(2\nu+1)(2x)^{-2\nu-1}$$

if $\nu>-\frac{1}{2}$, $x>0$. Also $\bar{u}(p)=\Gamma(\nu+1)p^{-\nu-1}$ and

$$\int_{-\infty}^{\infty}|\bar{u}(x+\mathrm{i}y)|^2\,\mathrm{d}y=\int_{-\infty}^{\infty}\frac{\Gamma^2(\nu+1)}{(x^2+y^2)^{\nu+1}}\mathrm{d}y$$

$$=2\Gamma^2(\nu+1)\int_0^{\infty}\frac{\mathrm{d}y}{(x^2+y^2)^{\nu+1}}.$$

The substitution $y=xs^{1/2}(1-s)^{-1/2}$ (i.e. $s=\sin^2(\arg p)$) now gives

$$\int_0^{\infty}\frac{\mathrm{d}y}{(x^2+y^2)^{\nu+1}}=\tfrac{1}{2}x^{-2\nu-1}\int_0^1 s^{-1/2}(1-s)^{\nu-1/2}\,\mathrm{d}s$$

$$=\frac{1}{2}\frac{\Gamma(\frac{1}{2})\Gamma(\nu+\frac{1}{2})}{\Gamma(\nu+1)}x^{-2\nu-1}.$$

The Parseval formula for $u(t)$ is therefore

$$\Gamma(\tfrac{1}{2})\Gamma(\nu+\tfrac{1}{2})\Gamma(\nu+1)x^{-2\nu-1}=2\pi\Gamma(2\nu+1)(2x)^{-2\nu-1},$$

which reduces to the duplication formula for the gamma function (E1.5.38).

P5.4.5 can easily be generalized as follows.

**P5.4.6** If $u(t)$ and $v(t)$ are real, $x>A$ and the integrals

$$\int_0^{\infty}u^2(t)\mathrm{e}^{-2At}\,\mathrm{d}t \qquad \text{and} \qquad \int_0^{\infty}v^2(t)\mathrm{e}^{-2At}\,\mathrm{d}t$$

converge, then

$$\int_{-\infty}^{\infty} \bar{u}(x+iy)\bar{v}(x-iy)\,dy = 2\pi \int_{0}^{\infty} u(t)v(t)e^{-2xt}\,dt.$$

In P5.4.5, replace $u(t)$ by $u(t) \pm v(t)$ and take the difference between the results. This gives

$$\int_{-\infty}^{\infty} [\bar{u}(x+iy)\bar{v}(x-iy) + \bar{u}(x-iy)\bar{v}(x+iy)]\,dy$$

$$= 4\pi \int_{0}^{\infty} u(t)v(t)e^{-2xt}\,dt.$$

But

$$\int_{-\infty}^{\infty} \bar{u}(x-iy)\bar{v}(x+iy)\,dy = \int_{-\infty}^{\infty} \bar{u}(x+iy_1)\bar{v}(x-iy_1)\,dy_1,$$

where $y_1 = -y$, so the result follows.

For complex-valued functions we have P5.4.7.

**P5.4.7** If

$$\int_{0}^{\infty} |w_1(t)|^2 e^{-2At}\,dt \qquad \text{and} \qquad \int_{0}^{\infty} |w_2(t)|^2 e^{-2At}\,dt$$

both converge and $x > A$, then

$$\int_{-\infty}^{\infty} \bar{w}_1(x+iy)\bar{w}_2(x-iy)\,dy = 2\pi \int_{0}^{\infty} w_1(t)w_2(t)e^{-2xt}\,dt.$$

Put $w_1(t) = u_1(t) + iv_1(t)$, $w_2(t) = u_2(t) + iv_2(t)$ and use P5.4.6.

If in P5.4.7 we take $w_2(t)$ to be the complex conjugate of $w_1(t)$, then $\bar{w}_2(x-iy)$ is the complex conjugate of $\bar{w}_1(x+iy)$.

**P5.4.8** If

$$\int_{0}^{\infty} |w(t)|^2 e^{-2At}\,dt$$

converges, then for $x > A$

$$\int_{-\infty}^{\infty} |\bar{w}(x+iy)|^2\,dy = 2\pi \int_{0}^{\infty} |w(t)|^2 e^{-2xt}\,dt.$$

P5.4.7 can be generalized further to P5.4.9.

**P5.4.9** If

$$\int_0^\infty |w_1(t)|^2 e^{-2A_1t}\,dt \qquad \text{and} \qquad \int_0^\infty |w_2(t)|^2 e^{-2A_2t}\,dt$$

both converge, and $\mathscr{R}z_1 > A_1$, $\mathscr{R}z_2 > A_2$, then

$$\int_{-\infty}^{\infty} \bar{w}_1(z_1 + iy)\bar{w}_2(z_2 - iy)\,dy$$
$$= 2\pi \int_0^\infty w_1(t)w_2(t)\exp[-(z_1 + z_2)t]\,dt.$$

Choose $x$ such that $0 < x < \min(\mathscr{R}z_1 - A_1, \mathscr{R}z_2 - A_2)$, and put $v_1(t) = w_1(t)\exp[(x - z_1)t]$, $v_2(t) = w_2(t)\exp[(x - z_2)t]$. The result then follows from P5.4.7 with $A = 0$.

The main interest of P5.4.9 is that it gives a formula for the Laplace transform of the product of two functions, on taking $p = z_1 + z_2$.

**P5.4.10** If $A_1$ and $A_2$ are as in P5.4.9 and $\mathscr{R}p > A_1 + A_2$, then

$$\mathscr{L}\{w_1(t)w_2(t)\} = \frac{1}{2\pi i}\int_{c-i\infty}^{c+i\infty} \bar{w}_1(z)\bar{w}_2(p - z)\,dz,$$

where the path of integration lies in the strip $A_1 < \mathscr{R}z < \mathscr{R}p - A_2$.

This formula is a parallel to the convolution property P1.2.2, since the right side is a complex convolution of the transforms $\bar{w}_1(p)$ and $\bar{w}_2(p)$. The result was used in P5.2.5.

## 5.5 Linear Operations on Transforms

The direct and inverse Laplace transformations are linear operations. Suppose that $T$ is a linear operator which acts on a function $u(t)$ defined for $t > 0$ so as to produce another function

$$v(t) = T[u(t)],$$

which is also defined for $t > 0$. Then provided the transforms $\bar{u}(p)$ and $\bar{v}(p)$ exist we can write

$$\bar{v}(p) = \bar{T}[\bar{u}(p)],$$

where $\bar{T}$ is another linear operator which acts on the transform $\bar{u}(p)$ to yield the transform $\bar{v}(p)$. The relation between the operators $T$ and $\bar{T}$ may be expressed as

$$\bar{T}[\mathscr{L}\{u(t)\}] = \mathscr{L}\{T[u(t)]\}$$

for all functions $u(t)$ for which both sides are defined. Similarly

$$T[\mathscr{L}^{-1}\{\bar{u}(p)\}] = \mathscr{L}^{-1}\{\bar{T}[\bar{u}(p)]\},$$

assuming that we can ignore null functions. Thus we have the operator equations

$$\bar{T} = \mathscr{L}T\mathscr{L}^{-1}, \qquad T = \mathscr{L}^{-1}\bar{T}\mathscr{L}.$$

Many results of this type have already been given. Examples include the change-of-scale property P1.1.2 and the translation properties P1.1.3 and P1.1.10. The convolution properties P1.2.2 and P5.4.10 are bilinear operator equations. Note that the integration property (P1.1.6)

$$v(t) = \int_0^t u(s)\mathrm{d}s \leftrightarrow \bar{v}(p) = p^{-1}\bar{u}(p)$$

holds for any function $u(t)$ with a transform $\bar{u}(p)$, but its inverse

$$u(t) = v'(t) \leftrightarrow \bar{u}(p) = p\bar{v}(p)$$

requires that $v(0) = 0$.

In this section we shall obtain other formulae of this type relating operators $T$ and $\bar{T}$. The calculations will be purely formal, and no attempt will be made to establish conditions for the validity of the results. In many cases the functions $u(t)$ or $\bar{u}(p)$ must be restricted in some way in order that $v(t)$ and $\bar{v}(p)$ may exist and satisfy the given equations (see E5.6.14).

The general formal results are found by applying the operators to the direct and inverse transforms under the sign of integration.

**P5.5.1** If $\bar{v}(p) = \bar{T}[\bar{u}(p)]$, then

$$v(t) = \int_0^\infty u(x)\phi(t, x)\mathrm{d}x,$$

where

$$\bar{\phi}(p, x) = \bar{T}[\mathrm{e}^{-px}].$$

For

$$\bar{v}(p) = \bar{T}\left[\int_0^\infty u(x)\mathrm{e}^{-px}\mathrm{d}x\right] = \int_0^\infty u(x)\bar{T}[\mathrm{e}^{-px}]\mathrm{d}x,$$

if we assume that the linear operator $\bar{T}$ can be interchanged with the integration. A similar assumption about the inverse Laplace transformation gives the result.

Similarly we obtain property P5.5.2.

**P5.5.2** If $v(t) = T[u(t)]$, then

$$\bar{v}(p) = \frac{1}{2\pi i} \int_{c-i\infty}^{c+i\infty} \bar{u}(z)\bar{\psi}(p, z)\,dz,$$

where

$$\psi(t, z) = T[e^{zt}],$$

**E5.5.1** $\bar{v}(p) = p^{-\nu-1}\bar{u}(p^{\lambda})$.

Here $\bar{\Phi}(p, x) = p^{-\nu-1}\exp(-xp^{\lambda})$, which is a Laplace transform when $x > 0$ provided that

either (i) $0 < \mathscr{R}\lambda < 1$, $\quad \nu$ arbitrary,
or (ii) $\mathscr{R}\nu > -1$, $\quad \mathscr{R}\lambda \leqslant 1$.

The case $\lambda = \frac{1}{2}$ is treated in E3.3.6. In particular (E1.3.10) P5.5.1 shows that $\bar{v}(p) = \bar{u}(p^{1/2})$ corresponds to

$$v(t) = \frac{1}{2\sqrt{(\pi t^3)}} \int_0^\infty u(x)\exp\left(-\frac{x^2}{4t}\right)x\,dx.$$

For $\lambda = -1$, $\mathscr{R}\nu > -1$, we find from E1.3.1 that

$$\bar{v}(p) = p^{-\nu-1}\bar{u}(p^{-1})$$

gives

$$v(t) = \int_0^\infty u(x)(t/x)^{\nu/2} J_\nu(2\sqrt{(xt)})\,dx.$$

The substitution $t = \frac{1}{2}s^2$, $x = \frac{1}{2}y^2$ leads to

$$s^{-\nu}v(\tfrac{1}{2}s^2) = \int_0^\infty y^{-\nu}u(\tfrac{1}{2}y^2)J_\nu(sy)y\,dy,$$

that is $s^{-\nu}v(\frac{1}{2}s^2)$ is the *Hankel transform* of $s^{-\nu}u(\frac{1}{2}s^2)$. The relation between $\bar{u}(p)$ and $\bar{v}(p)$ is symmetrical, so we infer that the Hankel transformation is also a symmetrical relation.

**E5.5.2** $v(t) = t^{\nu-1}u(t^{\lambda})$.

In this case $\psi(t, z) = t^{\nu-1}\exp(zt^{\lambda})$ has a Laplace transform for all $z$ when $0 \leqslant \mathscr{R}\lambda < 1$, $\mathscr{R}\nu > 0$. If $\lambda$ is real, the transform exists when $\mathscr{R}z < 0$, provided that $\mathscr{R}\nu > 0$ if $\lambda \geqslant 0$.

When $\lambda = \frac{1}{2}$, $\nu = 1$ we have $v(t) = u(t^{1/2})$ and

$$\bar{\psi}(p, z) = \int_0^\infty \exp(zt^{1/2} - pt)\mathrm{d}t$$

$$= \int_0^\infty 2s \exp(zs - ps^2)\mathrm{d}s$$

$$= p^{-1}\left(1 + z\int_0^\infty \exp(zs - ps^2)\mathrm{d}s\right).$$

Since

$$\int_{-\infty}^\infty \exp(zs - ps^2)\mathrm{d}s = \sqrt{(\pi/p)}\exp(z^2/4p),$$

we can write

$$\bar{\psi}(p, z) = \sqrt{\left(\frac{\pi}{p^3}\right)} z \exp\left(\frac{z^2}{4p}\right) + \frac{z}{p}\int_0^\infty [\exp(-zs) - \exp(-zs - ps^2)]\mathrm{d}s$$

for $\mathscr{R}z > 0$. If $\mathscr{R}p > 0$, the path of integration in the last integral can be taken in any direction such that $\mathscr{R}(ps^2) > 0$ and $\mathscr{R}(zs) > 0$, and it then follows that as $|z| \to \infty$ with $\mathscr{R}z > 0$

$$\frac{z}{p}\int_0^\infty \mathrm{e}^{-zs}[1 - \exp(-ps^2)]\mathrm{d}s \sim \frac{z}{p}\int_0^\infty ps^2\mathrm{e}^{-zs}\mathrm{d}s = \frac{2}{z^2}.$$

Thus when we substitute in P5.5.2 this term makes no contribution to $\bar{v}(p)$ and so

$$\bar{v}(p) = \frac{1}{2\sqrt{(\pi p^3)}\mathrm{i}}\int_{c-\mathrm{i}\infty}^{c+\mathrm{i}\infty} \bar{u}(z)\exp\left(\frac{z^2}{4p}\right) z\,\mathrm{d}z.$$

This result shows that

$$\mathscr{L}^{-1}\{\bar{u}(p^{1/2})\} = \frac{1}{4\sqrt{\pi}} t^{-3/2}\bar{v}(\tfrac{1}{4}t^{-1}),$$

and therefore

$$\bar{u}(p^{1/2}) = \int_0^\infty \frac{1}{4\sqrt{\pi}} t^{-3/2}\bar{v}(\tfrac{1}{4}t^{-1})\mathrm{e}^{-pt}\mathrm{d}p.$$

On interchanging $u$ and $v$ we have that if $v(t) = u(t^2)$, then

$$\bar{v}(p) = \int_0^\infty \tfrac{1}{2}(\pi z)^{-1/2}\bar{u}(z)\exp\left(-\frac{p^2}{4z}\right)\mathrm{d}z.$$

When $\lambda = -1$, $v(t) = t^{\nu-1}u(t^{-1})$ and

$$\psi(t, z) = t^{\nu-1}\exp(zt^{-1})$$

where $\mathscr{R}z < 0$. From E1.3.9

$$\bar{\psi}(p, z) = 2(-z/p)^{\nu/2}\,\mathrm{K}_\nu(2\sqrt{(-zp)}).$$

Since $\mathrm{K}_\nu(\zeta) \sim (\frac{1}{2}\pi/\zeta)^{1/2}\mathrm{e}^{-\zeta}$ we can modify the path of integration in P5.5.2 so that

$$\bar{v}(p) = \frac{1}{2\pi\mathrm{i}}\int_\infty^{(0-)} \bar{u}(z)\cdot 2\left(-\frac{z}{p}\right)^{\nu/2}\mathrm{K}_\nu(2\sqrt{(-zp)})\,\mathrm{d}z.$$

Here the $z$ plane is supposed to be cut from 0 to $\infty$ so that $zp \geqslant 0$ on the cut. If $|\bar{u}(z)|$ is suitably restricted as $z \to 0$ and $z \to \infty$, we can take the path of integration to coincide with the edges of the cut, and then

$$\bar{v}(p) = \int_0^\infty \bar{u}(z)\left(\frac{z}{p}\right)^{\nu/2}\mathrm{J}_\nu(2\sqrt{(zp)})\,\mathrm{d}z.$$

As in E5.5.1, this result may be interpreted as a Hankel transform.

## 5.6 Additional Examples

**E5.6.1** Solve the integral equations:

(i) $u(t) - 2\int_0^t u(s)\cos(t-s)\,\mathrm{d}s = t$;

(ii) $u(t) + \int_0^t u(s)(t-s+1)(t-s+3)\,\mathrm{d}s = g(t)$;

(iii) $\int_0^t u(s)\cos(t-s)\,\mathrm{d}s = \mathrm{t}$;

(iv) $\int_0^t u(s)(t-s)^{-1/2}\,\mathrm{d}s = \mathrm{erf}(\sqrt{t})$.

**E5.6.2** Solve the integro-differential equations:

(i) $u'(t) + 3u(t) = 5\int_0^t u(s)\sin[2(t-s)]\,\mathrm{d}s, \qquad u(0) = 1$;

(ii) $u'(t) + \int_0^t u(s)u(t-s)\,\mathrm{d}s = 0, \qquad u(0) = 1.$

**E5.6.3** Show that if

$$u'(t) = \int_0^t \frac{u(s)\,\mathrm{d}s}{\sqrt{[\pi(t-s)]}} \qquad \text{and} \qquad u(0) = 1$$

then

$$u(t) = \tfrac{2}{3}\mathrm{e}^t + \frac{1}{\pi}\int_0^\infty \frac{\sqrt{r}}{r^3+1}\mathrm{e}^{-rt}\,\mathrm{d}r,$$

and find the asymptotic expansion of $u(t)$ as $t \to \infty$.

**E5.6.4** The function $u(x, t)$ satisfies the equation

$$\frac{\partial u}{\partial x} + 2\frac{\partial u}{\partial t} = 1 + \int_0^t \frac{\partial u}{\partial t}(x, s)\,u(x, t-s)\,\mathrm{d}s,$$

and $u(x, 0) = u(0, t) = 0$. Show that

$$\bar{u}(x, p) = p^{-1}\{\sqrt{(p^2-1)}\coth[x\sqrt{(p^2-1)}] + p\}^{-1},$$

and deduce that

$$p\bar{u}(x, p) = \mathrm{e}^{-\zeta} - 2\sinh\zeta \sum_{n=1}^{\infty} \exp[-2n(x\sinh\zeta + \zeta)],$$

where $p = \cosh\zeta$. Hence obtain the expansion [compare E3.5.8]

$$\frac{\partial u}{\partial t} = t^{-1}\mathrm{I}_1(t) - \sum_{n=1}^{\infty} [\tfrac{1}{2}\sigma_n^{n-1}\mathrm{I}_{2n-2}(\tau_n) - \sigma_n^n \mathrm{I}_{2n}(\tau_n)$$
$$+ \tfrac{1}{2}\sigma_n^{n+1}\mathrm{I}_{2n+2}(\tau_n)]\,\mathrm{H}(t-2nx),$$

where

$$\sigma_n = \frac{t-2nx}{t+2nx}, \qquad \tau_n = \sqrt{(t^2 - 4n^2x^2)}.$$

Prove that $\bar{u}(x, p)$ is regular except for poles, and that as $t \to \infty$

$$u(x, t) \sim \begin{cases} \tan x + \dfrac{k^2-1}{k(kx+1)}\mathrm{e}^{kt} & (0 < x < \tfrac{1}{2}\pi), \\ t - \tfrac{1}{4}\pi & (x = \tfrac{1}{2}\pi), \\ \dfrac{k^2-1}{k(kx+1)}\mathrm{e}^{kt} + \tan x & (\tfrac{1}{2}\pi < x < \tfrac{3}{2}\pi). \end{cases}$$

Here $k$ is the real root of

$$k\tanh[x\sqrt{(k^2-1)}] = -\sqrt{(k^2-1)} \qquad \text{if } 0 < x < 1,$$

$k = -1$ if $x = 1$, and $k$ is the greatest real root of

$$k \tan[x\sqrt{(1-k^2)}] = -\sqrt{(1-k^2)} \qquad \text{if } x > 1.$$

[This problem comes from the theory of neutron transport (see Bellman *et al.*, 1964, pp. 23–26).]

**E5.6.5** Solve the difference equations:

(i) $u(t+2) + u(t+1) - 2u(t) = t, \qquad$ with $u(t) = 0$ for $0 < t < 2$;

(ii) $u(t+2) - 2u(t+1) + u(t) = e^{-t}$,
with $u(t) = 0$ $(0 < t < 1)$, $u(t) = 1$ $(1 < t < 2)$.

**E5.6.6** Solve the recurrence relations:

(i) $a_{n+2} = 2a_{n+1} + a_n, \qquad a_0 = a_1 = 1$;

(ii) $a_{n+2} + 2a_{n+1} + a_n = n, \qquad a_0 = 0, a_1 = 1$.

**E5.6.7** The function $u(t)$ satisfies the equation

$$u'(t) = u(t) - u(t-1)$$

and $u(t) = 0$ for $t < 0$, $u(0+) = 1$. Show that

$$\bar{u}(p) = (p - 1 + e^{-p})^{-1},$$

and deduce that

$$u(t) = \sum_{n=0}^{\infty} \frac{(-1)^n}{n!}(t-n)^n e^{t-n} H(t-n).$$

Prove that $\bar{u}(p)$ has a double pole at $p = 0$ and simple poles $p_n = x_n + iy_n$, $p_{-n} = x_n - iy_n$ where $x_n < 0$, $2n\pi < y_n < (2n + \frac{1}{2})\pi$ for $n = 1, 2, 3, \ldots$. Hence show that

$$u(t) = 2t + \tfrac{2}{3} - 2\sum_{n=1}^{\infty} \frac{x_n \cos y_n t + y_n \sin y_n t}{x_n^2 + y_n^2} \exp(x_n t)$$

where $x_n = \log(y_n^{-1} \sin y_n) = 1 - y_n \cot y_n$.

**E5.6.8** The function $u(t)$ satisfies the equation (Bellman and Kotkin, 1962)

$$tu'(t) + u(t-1) = 0$$

and $u(t) = 0$ for $t < 0$, $u(t) = 1$ for $0 < t \leqslant 1$. Show that

$$\bar{u}(p) = \exp[f(p)],$$

where

$$f(p) = \gamma - \int_0^p z^{-1}(1 - e^{-z})dz$$

and $\gamma$ is Euler's constant. Use the saddle point method to show that

$$u(t) \sim \left[2\pi\left(t - \frac{t-1}{\sigma}\right)\right]^{-1/2} \exp[-\sigma t + f(-\sigma)]$$

as $t \to \infty$, where $\sigma$ is the real root of $\sigma = \log(1 + \sigma t)$.

Show also that if the initial condition is changed to

$$u(t) = g(t) \qquad \text{for } -1 < t < 0,\ u(0+) = 1,$$

where $g(t) \to 0$ as $t \to -1$, then

$$\bar{u}(p) = A \exp[f(p) - \gamma] \log p + \bar{u}_1(p),$$

where

$$A = \int_{-1}^{0} g(s)ds$$

and $\bar{u}_1(p)$ is regular for all $p$. Deduce that in this case

$$u(t) = -A[t^{-1} + t^{-2} + \tfrac{3}{2}t^{-3} + O(t^{-4})] \qquad \text{as } t \to \infty.$$

**E5.6.9** The function $u(t)$ satisfies the equation

$$u'(t) = u(kt),$$

where $0 < k < 1$, and $u(0) = 1$. Show that

$$\bar{u}(p) = \sum_{n=0}^{\infty} k^{n(n-1)/2} p^{-n-1}$$

is regular for all $p \neq 0$, and that for all finite $t$

$$u(t) = \sum_{n=0}^{\infty} \frac{k^{n(n-1)/2}}{n!} t^n.$$

Prove that when $t > 0$

$$u(t) = \frac{1}{2\pi i} \int^{(0+)} \sum_{n=-\infty}^{\infty} k^{n(n-1)/2} p^{-n-1} e^{pt} dp,$$

put $p = e^z$ and $k = e^{-2/h}$, and use E3.5.6 to obtain

$$u(t) = \frac{\sqrt{(\pi h)}}{2\pi i} \int_{c-i\infty}^{c+i\infty} \exp[te^z + \tfrac{1}{4}h(z - h^{-1})^2] dz.$$

Apply the saddle point method to this integral to show that as $t \to \infty$

$$u(t) \sim (\xi + 1)^{-1/2} \exp\left[\tfrac{1}{4}h(\xi^2 + 2\xi)\right],$$

where $\xi$ is the real root of

$$\xi e^{\xi} = \frac{2}{h} e^{1/h} t.$$

**E5.6.10** Use P5.4.6 and E1.3.9 to show that

$$\int_{-\infty}^{\infty} \mathrm{K}_0(2\sqrt{[a(x+\mathrm{i}y)]})\mathrm{K}_0(2\sqrt{[b(x-\mathrm{i}y)]})\mathrm{d}y$$
$$= \pi \sqrt{\left(\frac{2x}{a+b}\right)} \mathrm{K}_1(2\sqrt{[2(a+b)x]}),$$

where $a$, $b$ and $x$ are positive. By making $x \to 0$, show that

$$\int_0^{\infty} [\ker(2\sqrt{(ay)})\ker(2\sqrt{(by)})$$
$$+ \operatorname{kei}(2\sqrt{(ay)})\operatorname{kei}(2\sqrt{(by)})]\mathrm{d}y = \frac{\pi}{4(a+b)},$$

where $\mathrm{K}_0(z\sqrt{(\pm \mathrm{i})}) = \ker(z) \pm \mathrm{i}\operatorname{kei}(z)$.

**E5.6.11** Prove that the functions

$$u_n(t) = \mathrm{H}(t-n) - \mathrm{H}(t-n-1) \qquad (n = 0, 1, 2, \dots)$$

are orthogonal and normal over the range $0 < t < \infty$. Deduce from P5.4.6 that the functions $\sqrt{(2/\pi)}\mathrm{e}^{\mathrm{i}ny}\sin(\frac{1}{2}y)y^{-1}$ are orthogonal and normal over $-\infty < y < \infty$.

**E5.6.12** Obtain the formulae:

(i) $$p^{-\nu-1}\bar{u}(p + p^{-1})$$
$$= \mathscr{L}\left\{\int_0^t u(x)\left(\frac{t-x}{x}\right)^{\nu/2} \mathrm{J}_\nu(2\sqrt{[x(t-x)]})\mathrm{d}x\right\} \qquad \text{for } \mathscr{R}\nu > -1;$$

(ii) $$p^{-\nu}\bar{u}(\log p) = \mathscr{L}\left\{\int_0^{\infty} \frac{t^{x+\nu-1}}{\Gamma(x+\nu)} u(x)\mathrm{d}x\right\} \qquad \text{for } \mathscr{R}\nu \geqslant 0;$$

(iii) $$\bar{u}(\sqrt{(p^2-\alpha^2)}) - \bar{u}(p) = \mathscr{L}\left\{\alpha \int_0^t u(\sqrt{(t^2-s^2)})\mathrm{I}_1(\alpha s)\mathrm{d}s\right\}.$$

**E5.6.13** Let $v(t) = u(n+\frac{1}{2})$ when $n < t < n+1$ for $n = 0, 1, 2, \ldots$. Obtain the formula

$$\bar{v}(p) = \frac{1 - e^{-p}}{2\pi i p} \int_{c-i\infty}^{c+i\infty} \frac{e^{z/2}\bar{u}(z)}{1 - e^{z-p}} dz,$$

where $\mathscr{R}z < \mathscr{R}p$ on the path of integration. Deduce that

$$\bar{v}(p) = 2p^{-1} \sinh \tfrac{1}{2}p \sum_{n=-\infty}^{\infty} (-1)^n \bar{u}(p + 2n\pi i).$$

**E5.6.14** Show that if the appropriate double integrals are absolutely convergent, then

(i) $\mathscr{L}\left\{\int_0^\infty \frac{t^{x-1}}{\Gamma(x)} u(x) dx\right\} = \bar{u}(\log p);$

(ii) $\frac{1}{\Gamma(p+1)} \int_0^\infty \bar{u}(x) x^p e^{-x} dx = \mathscr{L}\{u(e^t - 1)\}.$

# References

Numbers in square brackets are references to pages in the text.


R. Bellman, H. Kagiwada, R. Kalaba and M. Prestrud (1964). *Invariant Imbedding and Time-Dependent Processes.* American Elsevier. [185]

R. Bellman and B. Kotkin (1962). On the numerical solution of a differential-difference equation occurring in analytic number theory. *Math. Comp.* **16** 473–475. [185]

H.S. Carslaw and J.C. Jaeger (1947). *Conduction of Heat in Solids.* Oxford University Press. [120]

G. Doetsch (1974). *Introduction to the Theory and Application of the Laplace Transformation.* Springer. [vii]

A. Erdélyi (Ed.) (1954). *Tables of Integral Transforms*, Vol. I, McGraw-Hill. [viii]

A. Erdélyi (1962). *Operational Calculus and Generalized Functions.* Holt, Rinehart and Winston. [vii]

E.L. Ince (1927). *Ordinary Differential Equations.* Longman. [50]

M.J. Lighthill (1950). Contributions to the theory of heat transfer through a laminar boundary layer. *Proc. R. Soc.* A **202** 359–377. [139]

A.E.H. Love (1927). *A Treatise on the Mathematical Theory of Elasticity.* Cambridge University Press. [135]

L.M. Milne-Thomson (1933). *The Calculus of Finite Differences.* Macmillan. [161]

F.W.J. Olver (1974). *Asymptotics and Special Functions.* Academic Press. [98, 143]

L.J. Slater (1960). *Confluent Hypergeometric Functions.* Cambridge University Press. [33]

K. Stewartson (1957). On asymptotic expansions in the theory of boundary layers. *J. Math. Phys.* **36** 173–191. [141]

A. Talbot (1979). The accurate numerical inversion of Laplace transforms, *J. Inst. Maths Applics* **23** 97–120. [102]

E.C. Titchmarsh (1937). *Introduction to the Theory of Fourier Integrals.* Oxford University Press. [vii]

G.N. Ward (1955). *Linearized Theory of Steady High-Speed Flow.* Cambridge University Press. [119]

G.N. Watson (1944). *Theory of Bessel Functions.* Cambridge University Press. [19, 22, 26, 98, 120, 127, 128, 143]

D.V. Widder (1941). *The Laplace Transform.* Princeton University Press. [vii]

# List of Transforms

## 1 General Properties

| | $\mathscr{L}\{u(t)\} = \bar{u}(p)$ | $\mathscr{L}^{-1}\{\bar{u}(p)\} = u(t)$ | |
|---|---|---|---|
| **1.1** | $\int_0^\infty u(t)e^{-pt}\,dt$ | $u(t)$ | D1.1.1 |
| **1.2** | $\bar{u}(p)$ | $\frac{1}{2\pi i}\int_{c-i\infty}^{c+i\infty} \bar{u}(p)e^{pt}\,dp$ | P3.2.7 |
| **1.3** | $\lambda\bar{u}(p) + \mu\bar{v}(p)$ | $\lambda u(t) + \mu v(t)$ | P1.1.1 |
| **1.4** | $\bar{u}(p/k)$ | $ku(kt) \quad (k > 0)$ | P1.1.2 |
| **1.5** | $\bar{u}(p - \alpha)$ | $e^{\alpha t}u(t)$ | P1.1.3 |
| **1.6** | $p\bar{u}(p) - u(0)$ | $u'(t)$ | P1.1.4 |
| **1.7** | $p^2\bar{u}(p) - pu(0) - u'(0)$ | $u''(t)$ | P1.1.5 |
| **1.8** | $p^n\bar{u}(p) - p^{n-1}u(0) - \dots - u^{(n-1)}(0)$ | $u^{(n)}(t)$ | P1.1.5 |
| **1.9** | $p^{-1}\bar{u}(p)$ | $\int_0^t u(s)\,ds$ | P1.1.6 |
| **1.10** | $e^{-ap}\bar{u}(p)$ | $u(t-a)H(t-a) \quad (a > 0)$ | P1.1.10 |
| **1.11** | $e^{ap}\left(\bar{u}(p) - \int_0^a u(t)e^{-pt}\,dt\right)$ | $u(t+a) \quad (a > 0)$ | P5.2.1 |
| **1.12** | $(1 - e^{-pT})\bar{v}(p)$ | $\begin{cases} u(t) = u(t-T) \\ v(t) = u(t)H(T-t) \end{cases} \quad (T > 0)$ | P1.1.11 |
| **1.13** | $\bar{u}'(p)$ | $-tu(t)$ | P1.1.9, P3.1.1 |
| **1.14** | $\int_p^\infty \bar{u}(z)\,dz$ | $t^{-1}u(t)$ | P1.3.3 |
| **1.15** | $\frac{\partial}{\partial\lambda}\bar{u}(p,\lambda)$ | $\frac{\partial}{\partial\lambda}u(t,\lambda)$ | P1.3.2 |
| **1.16** | $\bar{u}(p)\bar{v}(p)$ | $\int_0^t u(s)v(t-s)\,ds$ | P1.2.2 |
| **1.17** | $\frac{1}{2\pi i}\int_{c-i\infty}^{c+i\infty} \bar{u}(z)\bar{v}(p-z)\,dz$ | $u(t)v(t)$ | P5.4.10 |

| | | | |
|---|---|---|---|
| **1.18** | $\int_0^\infty u(s)\bar{v}(p+s)\mathrm{d}s$ | $\bar{u}(t)v(t)$ | E1.5.33 |
| **1.19** | $\int_p^\infty (z-p)^{\nu-1}\bar{u}(z)\mathrm{d}z$ | $\Gamma(\nu)t^{-\nu}u(t) \quad (\mathscr{R}\nu>0)$ | E1.5.33 |
| **1.20** | $\int_0^\infty \frac{u(s)}{p+s}\mathrm{d}s$ | $\bar{u}(t)$ | E1.5.33 |
| **1.21** | $\bar{u}(p^{1/2})$ | $\int_0^\infty u(s)\frac{s}{2\sqrt{(\pi t^3)}}\exp\left(-\frac{s^2}{4t}\right)\mathrm{d}s$ | E5.5.1 |
| **1.22** | $p^{-\nu-1}\bar{u}(p^{-1})$ | $\int_0^\infty u(s)\left(\frac{t}{s}\right)^{\nu/2}\mathrm{J}_\nu(2\sqrt{(ts)})\mathrm{d}s$ | E5.5.1 |
| **1.23** | $p^{-\nu-1}\bar{u}(p+p^{-1})$ | $\int_0^t u(s)\left(\frac{t-s}{s}\right)^{\nu/2}$ $\times \mathrm{J}_\nu(2\sqrt{[s(t-s)]})\mathrm{d}s$ | E5.6.12 |
| **1.24** | $p^{-\nu}\bar{u}(\log p)$ | $\int_0^\infty \frac{t^{s+\nu-1}}{\Gamma(s+\nu)}u(s)\mathrm{d}s$ | E5.6.12 |
| **1.25** | $\bar{u}(\sqrt{(p^2-\alpha^2)})$ | $\alpha\int_0^t u(\sqrt{(t^2-s^2)})\mathrm{I}_1(\alpha s)\mathrm{d}s$ | E5.6.12 |
| **1.26** | $\frac{1}{2\sqrt{(\pi p^3)}\mathrm{i}} \times \int_{c-\mathrm{i}\infty}^{c+\mathrm{i}\infty} \bar{u}(z)\exp\left(\frac{z^2}{4p}\right)z\,\mathrm{d}z$ | $u(t^{1/2})$ | E5.5.2 |
| **1.27** | $\frac{1}{2\sqrt{\pi}}\int_0^\infty \bar{u}(x)x^{-1/2}\exp\left(-\frac{p^2}{4x}\right)\mathrm{d}x$ | $u(t^2)$ | E5.5.2 |
| **1.28** | $\int_0^\infty \bar{u}(x)\left(\frac{x}{p}\right)^{\nu/2}\mathrm{J}_\nu(2\sqrt{(px)})\mathrm{d}x$ | $t^{\nu-1}u(t^{-1})$ | E5.5.2 |
| **1.29** | $\frac{1}{\Gamma(p+1)}\int_0^\infty \bar{u}(x)x^p\mathrm{e}^{-x}\mathrm{d}x$ | $u(\mathrm{e}^t-1)$ | E5.6.14 |

## 2 Special Functions

| | $\bar{u}(p)$ | $u(t)$ | |
|---|---|---|---|
| **2.1** | $p^{-1}$ | $1$ | E1.1.1 |
| **2.2** | $p^{-2}$ | $t$ | E1.1.2 |
| **2.3** | $p^{-n-1}$ | $t^n/n!$ | E1.1.6 |
| **2.4** | $(p-\alpha)^{-1}$ | $\mathrm{e}^{\alpha t}$ | E1.1.3 |
| **2.5** | $p/(p^2+\alpha^2)$ | $\cos\alpha t$ | E1.1.4 |

| | | | |
|---|---|---|---|
| **2.6** | $\dfrac{\alpha}{p^2+\alpha^2}$ | $\sin \alpha t$ | E1.1.5 |
| **2.7** | $\dfrac{p}{p^2-\alpha^2}$ | $\cosh \alpha t$ | |
| **2.8** | $\dfrac{\alpha}{p^2-\alpha^2}$ | $\sinh \alpha t$ | |
| **2.9** | $\dfrac{\alpha^3}{(p^2+\alpha^2)^2}$ | $\frac{1}{2}\sin \alpha t - \frac{1}{2}\alpha t \cos \alpha t$ | E1.4.3 |
| **2.10** | $\dfrac{\alpha^2 p}{(p^2+\alpha^2)^2}$ | $\frac{1}{2}\alpha t \sin \alpha t$ | E1.4.4 |
| **2.11** | $\dfrac{\alpha p^2}{(p^2+\alpha^2)^2}$ | $\frac{1}{2}\alpha t \cos \alpha t + \frac{1}{2}\sin \alpha t$ | |
| **2.12** | $\dfrac{p^3}{(p^2+\alpha^2)^2}$ | $\cos \alpha t - \frac{1}{2}\alpha t \sin \alpha t$ | |
| **2.13** | $\dfrac{4\alpha^3}{p^4+4\alpha^4}$ | $\cosh \alpha t \sin \alpha t - \sinh \alpha t \cos \alpha t$ | E1.5.3 |
| **2.14** | $\dfrac{2\alpha^2 p}{p^4+4\alpha^4}$ | $\sinh \alpha t \sin \alpha t$ | E1.5.1(vi) |
| **2.15** | $\dfrac{2\alpha p^2}{p^4+4\alpha^4}$ | $\cosh \alpha t \sin \alpha t + \sinh \alpha t \cos \alpha t$ | |
| **2.16** | $\dfrac{p^3}{p^4+4\alpha^4}$ | $\cosh \alpha t \cos \alpha t$ | E1.5.3 |
| **2.17** | $\dfrac{\alpha^n}{p(p+\alpha)\dots(p+n\alpha)}$ | $\dfrac{(1-e^{-\alpha t})^n}{n!}$ | E1.5.5 |
| **2.18** | $(p-1)^n p^{-n-1}$ | $L_n(t)$ | E1.5.4 |
| **2.19** | $p^{-\nu}$ | $\dfrac{t^{\nu-1}}{\Gamma(\nu)} \quad (\mathscr{R}\nu > 0)$ | E1.2.2 |
| **2.20** | $\sqrt{(p-\alpha)} - \sqrt{(p-\beta)}$ | $\dfrac{e^{\beta t}-e^{\alpha t}}{2\sqrt{(\pi t^3)}}$ | |
| **2.21** | $(p^2+\alpha^2)^{-1/2}$ | $J_0(\alpha t)$ | E1.3.2 |
| **2.22** | $(p^2-\alpha^2)^{-1/2}$ | $I_0(\alpha t)$ | |
| **2.23** | $(p^2+\alpha^2)^{-\nu}$ | $\dfrac{\sqrt{\pi}}{\Gamma(\nu)}\left(\dfrac{t}{2\alpha}\right)^{\nu-1/2} J_{\nu-1/2}(\alpha t)$ $(\mathscr{R}\nu > 0)$ | |
| **2.24** | $(p^2-\alpha^2)^{-\nu}$ | $\dfrac{\sqrt{\pi}}{\Gamma(\nu)}\left(\dfrac{t}{2\alpha}\right)^{\nu-1/2} I_{\nu-1/2}(\alpha t)$ $(\mathscr{R}\nu > 0)$ | E1.5.17 |

| | | | |
|---|---|---|---|
| **2.25** | $\dfrac{[\sqrt{(p^2+\alpha^2)}+p]^{-\nu}}{\sqrt{(p^2+\alpha^2)}}$ | $\alpha^{-\nu}\mathrm{J}_\nu(\alpha t) \qquad (\mathscr{R}\nu > -1)$ | E2.4.3 |
| **2.26** | $\dfrac{[\sqrt{(p^2-\alpha^2)}+p]^{-\nu}}{\sqrt{(p^2-\alpha^2)}}$ | $\alpha^{-\nu}\mathrm{I}_\nu(\alpha t) \qquad (\mathscr{R}\nu > -1)$ | |
| **2.27** | $[\sqrt{(p^2+\alpha^2)}+p]^{-\nu}$ | $\dfrac{\nu}{t}\alpha^{-\nu}\mathrm{J}_\nu(\alpha t) \qquad (\mathscr{R}\nu > 0)$ | |
| **2.28** | $[\sqrt{(p^2-\alpha^2)}+p]^{-\nu}$ | $\dfrac{\nu}{t}\alpha^{-\nu}\mathrm{I}_\nu(\alpha t) \qquad (\mathscr{R}\nu > 0)$ | |
| **2.29** | $p^{-1}(p+\alpha)^{-1/2}$ | $\alpha^{-1/2}\,\mathrm{erf}(\sqrt{(\alpha t)})$ | E1.5.26 |
| **2.30** | $p^{-1}\sqrt{(p+\alpha)}$ | $\dfrac{\mathrm{e}^{-\alpha t}}{\sqrt{(\pi t)}}+\sqrt{(\alpha)}\,\mathrm{erf}(\sqrt{(\alpha t)})$ | E1.5.26 |
| **2.31** | $\dfrac{1}{\sqrt{p}+\sqrt{\alpha}}$ | $\dfrac{1}{\sqrt{(\pi t)}}-\sqrt{(\alpha)}\,\mathrm{e}^{\alpha t}\,\mathrm{erfc}(\sqrt{(\alpha t)})$ | E1.5.26 |
| **2.32** | $\dfrac{1}{p+\sqrt{(\alpha p)}}$ | $\mathrm{e}^{\alpha t}\,\mathrm{erfc}(\sqrt{(\alpha t)})$ | |
| **2.33** | $p^{a-c}(p-\alpha)^{-a}$ | $\dfrac{t^{c-1}}{\Gamma(c)}\,{}_1F_1(a;c;\alpha t) \qquad (\mathscr{R}c > 0)$ | E1.5.18 |
| **2.34** | $\exp[-2\sqrt{(\alpha p)}]$ | $\sqrt{\left(\dfrac{\alpha}{\pi}\right)}t^{-3/2}\exp\left(-\dfrac{\alpha}{t}\right)$ $(\mathscr{R}\alpha > 0)$ | E1.3.10 |
| **2.35** | $p^{-1/2}\exp[-2\sqrt{(\alpha p)}]$ | $\dfrac{1}{\sqrt{(\pi t)}}\exp\left(-\dfrac{\alpha}{t}\right) \qquad (\mathscr{R}\alpha > 0)$ | E1.3.10 |
| **2.36** | $p^{(n-1)/2}\exp[-2\sqrt{(\alpha p)}]$ | $\sqrt{\left(\dfrac{2}{\pi}\right)}(2t)^{-(n-1)/2}h_n\left(\sqrt{\left(\dfrac{2\alpha}{t}\right)}\right) \times \exp\left(-\dfrac{\alpha}{t}\right) \qquad (\mathscr{R}\alpha > 0)$ | E3.3.6 |
| **2.37** | $p^{-1}\exp[-2\sqrt{(\alpha p)}]$ | $\mathrm{erfc}\left(\sqrt{\dfrac{\alpha}{t}}\right) \qquad (\mathscr{R}\alpha > 0)$ | E3.3.6 |
| **2.38** | $p^{-1-n/2}\exp[-2\sqrt{(\alpha p)}]$ | $(4t)^{n/2}\,\mathrm{i}^n\,\mathrm{erfc}\left(\sqrt{\dfrac{\alpha}{t}}\right) \qquad (\mathscr{R}\alpha > 0)$ | E3.3.6 |
| **2.39** | $p^{-3/4}\exp(-x\sqrt{p})$ | $\dfrac{1}{\pi}\left(\dfrac{x}{2t}\right)^{1/2}\exp\left(-\dfrac{x^2}{8t}\right)\mathrm{K}_{1/4}\left(\dfrac{x^2}{8t}\right)$ $(\lvert\arg x\rvert < \tfrac{1}{4}\pi)$ | E3.3.6 |
| **2.40** | $p^{-1/2}\exp[-\sqrt{(ap)}]\cos[\sqrt{(ap)}]$ | $(\pi t)^{-1/2}\cos\dfrac{a}{2t} \qquad (a > 0)$ | E1.3.11 |
| **2.41** | $p^{-1/2}\exp[-\sqrt{(ap)}]\sin[\sqrt{(ap)}]$ | $(\pi t)^{-1/2}\sin\dfrac{a}{2t} \qquad (a > 0)$ | E1.3.11 |

| | | | |
|---|---|---|---|
| **2.42** | $\dfrac{\exp(-2\alpha\sqrt{p})}{p+\beta\sqrt{p}}$ | $\exp(2\alpha\beta+\beta^2 t)\operatorname{erfc}\left(\dfrac{\alpha}{\sqrt{t}}+\beta\sqrt{t}\right)$ <br> $(\lvert\arg\alpha\rvert<\tfrac{1}{4}\pi)$ | E1.5.29 |
| **2.43** | $p^{-1}\exp[-2\alpha\sqrt{(p+\beta^2)}]$ <br> $(\lvert\arg\alpha\rvert<\tfrac{1}{4}\pi)$ | $\tfrac{1}{2}e^{2\alpha\beta}\operatorname{erfc}\left(\dfrac{\alpha}{\sqrt{t}}+\beta\sqrt{t}\right)$ <br> $+\tfrac{1}{2}e^{-2\alpha\beta}\quad\operatorname{erfc}\left(\dfrac{\alpha}{\sqrt{t}}-\beta\sqrt{t}\right)$ | E1.5.28 |
| **2.44** | $p^{-\nu-1}e^{-\alpha/p}$ | $\left(\dfrac{t}{\alpha}\right)^{\nu/2}J_\nu(2\sqrt{(\alpha t)}\quad(\mathscr{R}\nu>-1)$ | E1.3.1 |
| **2.45** | $p^{-1/2}e^{-\alpha/p}$ | $(\pi t)^{-1/2}\cos[2\sqrt{(\alpha t)}]$ | |
| **2.46** | $p^{-3/2}e^{-\alpha/p}$ | $(\pi\alpha)^{-1/2}\sin[2\sqrt{(\alpha t)}]$ | |
| **2.47** | $p^{-\nu-1}e^{\alpha/p}$ | $\left(\dfrac{t}{\alpha}\right)^{\nu/2}I_\nu(2\sqrt{(\alpha t)}\quad(\mathscr{R}\nu>-1)$ | |
| **2.48** | $\dfrac{\exp[-a\sqrt{(p^2-\kappa^2)}]}{\sqrt{(p^2-\kappa^2)}}$ | $I_0(\kappa\sqrt{(t^2-a^2)})H(t-a)\ (a>0)$ | E3.5.8 |
| **2.49** | $\dfrac{\exp[\alpha p-\alpha\sqrt{(p^2-\kappa^2)}]}{\sqrt{(p^2-\kappa^2)}}$ | $J_0(\kappa\sqrt{[t(t+2\alpha)]})$ | |
| **2.50** | $\exp[-a\sqrt{(p^2-\kappa^2)}]-e^{-ap}$ | $\dfrac{\kappa a}{\sqrt{(t^2-a^2)}}I_1(\kappa\sqrt{(t^2-a^2)})$ <br> $\times H(t-a)\qquad(a>0)$ | E4.4.1 |
| **2.51** | $\left(\dfrac{\kappa}{p+\sqrt{(p^2-\kappa^2)}}\right)^\nu$ <br> $\times\dfrac{\exp[-a\sqrt{(p^2-\kappa^2)}]}{\sqrt{(p^2-\kappa^2)}}$ | $\left(\dfrac{t-a}{t+a}\right)^{\nu/2}I_\nu(\kappa\sqrt{(t^2-a^2)})$ <br> $\times H(t-a)\qquad(a>0)$ | E5.6.4 |
| **2.52** | $\dfrac{\cosh[(1-\alpha)\sqrt{p}]}{\sqrt{(p)}\sinh(\sqrt{p})}$ | $\vartheta_3(\tfrac{1}{2}\alpha\,\vert\,i\pi t)$ <br> $(\lvert\arg\alpha\rvert<\tfrac{1}{4}\pi,\lvert\arg(2-\alpha)\rvert<\tfrac{1}{4}\pi)$ | E3.5.6 |
| **2.53** | $p^{-1}\log p$ | $-\log t-\gamma$ | E1.3.3 |
| **2.54** | $p^{-\nu}\log p$ | $\dfrac{t^{\nu-1}}{\Gamma(\nu)}[\psi(\nu)-\log t]\qquad(\mathscr{R}\nu>0)$ | E1.3.3 |
| **2.55** | $\log\dfrac{p-\alpha}{p-\beta}$ | $\dfrac{1}{t}(e^{\beta t}-e^{\alpha t})$ | E1.5.16 |
| **2.56** | $\log\dfrac{p^2+\alpha^2}{p^2+\beta^2}$ | $\dfrac{2}{t}(\cos\beta t-\cos\alpha t)$ | E1.3.5 |
| **2.57** | $p^{-1}\log(p+\alpha)$ | $\log\alpha+E_1(\alpha t)$ | E1.5.24 |

| | | | |
|---|---|---|---|
| **2.58** | $\sinh^{-1}\sqrt{\dfrac{\alpha}{p}}$ | $\dfrac{1}{2t}\operatorname{erf}(\sqrt{(\alpha t)})$ | E1.4.8 |
| **2.59** | $\tan^{-1}\left(\dfrac{\alpha}{p}\right)$ | $t^{-1}\sin\alpha t$ | E1.3.4 |
| **2.60** | $(p^2+\alpha^2)^{-1/2}\sinh^{-1}\left(\dfrac{p}{\alpha}\right)$ | $-\frac{1}{2}\pi\,\mathrm{Y}_0(\alpha t)$ | E2.4.3 |
| **2.61** | $(p^2-\alpha^2)^{-1/2}\cosh^{-1}\left(\dfrac{p}{\alpha}\right)$ | $\mathrm{K}_0(\alpha t)$ | |
| **2.62** | $\dfrac{\Gamma(p/\alpha)\Gamma(\nu)}{\Gamma\left(\dfrac{p}{\alpha}+\nu\right)}$ | $\alpha(1-\mathrm{e}^{-\alpha t})^{\nu-1}$ $(\mathscr{R}\nu>0,\ \mathscr{R}\alpha>0)$ | E1.5.15 |
| **2.63** | $\psi\left(\dfrac{p}{\alpha}\right)-\log\dfrac{p}{\alpha}$ | $\dfrac{1}{t}-\dfrac{\alpha}{1-\mathrm{e}^{-\alpha t}}\quad(\mathscr{R}\alpha>0)$ | E1.5.22 |
| **2.64** | $p^{-1}\psi\left(\dfrac{p}{\alpha}\right)$ | $-\log(\mathrm{e}^{\alpha t}-1)-\gamma\quad(\mathscr{R}\alpha>0)$ | E1.5.22 |
| **2.65** | $\psi\left(\dfrac{p}{\alpha}\right)$ | $\dfrac{\alpha^2 t}{1-\mathrm{e}^{-\alpha t}}\quad(\mathscr{R}\alpha>0)$ | E1.5.22 |
| **2.66** | $\mathrm{e}^{\alpha p}E_1(\alpha p)$ | $(t+\alpha)^{-1}$ ($\alpha>0$ or not real) | E1.3.7 |
| **2.67** | $[\frac{1}{2}\pi-\mathrm{Si}(ap)]\cos ap+\mathrm{Ci}(ap)\sin ap$ | $\dfrac{a}{t^2+a^2}\quad(a>0)$ | E1.3.8 |
| **2.68** | $\exp\left(\dfrac{p^2}{4\alpha^2}\right)\operatorname{erfc}\left(\dfrac{p}{2\alpha}\right)$ | $\dfrac{2}{\sqrt{\pi}}\alpha\exp(-\alpha^2t^2)$ $(\lvert\arg\alpha\rvert\leqslant\frac{1}{4}\pi)$ | E1.3.6 |
| **2.69** | $\displaystyle\int_p^\infty \sin(z^2-p^2)\,\mathrm{d}z$ | $\frac{1}{2}\cos(\frac{1}{4}t^2)$ | E2.5.13 |
| **2.70** | $\displaystyle\int_p^\infty \cos(z^2-p^2)\,\mathrm{d}z$ | $\frac{1}{2}\sin(\frac{1}{4}t^2)$ | E2.5.13 |
| **2.71** | $p^{-1/2}\exp\left(-\dfrac{\alpha^2}{4p}\right)\operatorname{erfc}\left(\dfrac{\alpha}{2\sqrt{p}}\right)$ | $(\pi t)^{-1/2}\exp(-\alpha\sqrt{t})$ | |
| **2.72** | $\mathrm{Hi}(-p)$ | $\pi^{-1}\exp(-\frac{1}{3}t^3)$ | D1.5.5 |
| **2.73** | $E_n(p)$ | $t^{-n}\mathrm{H}(t-1)$ | D1.3.4 |
| **2.74** | $\mathrm{K}_0(ap)$ | $(t^2-a^2)^{-1/2}\mathrm{H}(t-a)\quad(a>0)$ | E4.2.3 |
| **2.75** | $\mathrm{K}_\nu(ap)$ | $\dfrac{\cosh[\nu\cosh^{-1}(t/a)]}{\sqrt{(t^2-a^2)}}\mathrm{H}(t-a)$ $(a>0)$ | |

| | | | |
|---|---|---|---|
| **2.76** | $\left(\frac{\alpha}{p}\right)^{\nu/2} \mathrm{K}_\nu(2\sqrt{(\alpha p)})$ | $\frac{1}{2}t^{\nu-1}\exp\left(-\frac{\alpha}{t}\right) \quad (\mathscr{R}\alpha > 0)$ | E1.3.9 |
| **2.77** | $p^{-1/2}\mathrm{K}_\nu(2\sqrt{(2\alpha p)})$ | $\frac{1}{2\sqrt{\pi}}t^{-3/2}\exp\left(-\frac{\alpha}{t}\right)\mathrm{K}_{\nu/2}\left(\frac{\alpha}{t}\right)$ $(\mathscr{R}\alpha > 0)$ | E3.5.12 |
| **2.78** | $p^{-1/2}\exp\left(\frac{\alpha}{2p}\right)\mathrm{I}_{\nu/2}\left(\frac{\alpha}{2p}\right)$ | $(\pi t)^{-1/2}\mathrm{I}_\nu(2\sqrt{(\alpha t)}) \quad (\mathscr{R}\nu > -1)$ | E1.5.21 |
| **2.79** | $p^{-1/2}\exp\left(-\frac{\alpha}{2p}\right)\mathrm{I}_{\nu/2}\left(\frac{\alpha}{2p}\right)$ | $(\pi t)^{-1/2}\mathrm{J}_\nu(2\sqrt{(\alpha t)}) \quad (\mathscr{R}\nu > -1)$ | |
| **2.80** | $p^{-1}\exp\left(-\frac{\alpha+\beta}{p}\right)\mathrm{I}_\nu\left(\frac{2\sqrt{(\alpha\beta)}}{p}\right)$ | $\mathrm{J}_\nu(2\sqrt{(\alpha t)})\mathrm{J}_\nu(2\sqrt{(\beta t)})$ $(\mathscr{R}\nu > -1)$ | E1.5.25 |
| **2.81** | $\mathrm{I}_\nu(x\sqrt{p})\mathrm{K}_\nu(y\sqrt{p})$ | $\frac{1}{2t}\exp\left(-\frac{x^2+y^2}{4t}\right)\mathrm{I}_\nu\left(\frac{xy}{2t}\right)$ $(0 < x \leqslant y)$ | E3.5.13 |
| **2.82** | $p^{-b}F\left(a, b; c; \frac{\alpha}{p}\right)$ | $\frac{t^{b-1}}{\Gamma(b)}{}_1F_1(a; c; \alpha t) \quad (\mathscr{R}b > 0)$ | E1.5.18 |
| **2.83** | ${}_1F_1(a; c; -p)$ | $\frac{\Gamma(c)}{\Gamma(a)\Gamma(c-a)}t^{a-1}(1-t)^{c-a-1}$ $\times\, \mathrm{H}(1-t) \quad (\mathscr{R}c > \mathscr{R}a > 0)$ | E1.5.20 |
| **2.84** | $U(a; c; p)$ | $\frac{t^{a-1}(1+t)^{c-a-1}}{\Gamma(a)} \quad (\mathscr{R}a > 0)$ | E2.5.15 |
| **2.85** | $p^{-1}\tanh(\frac{1}{4}pT)$ | $\begin{cases} 1 & (nT < t < (n+\frac{1}{2})T) \\ -1 & ((n+\frac{1}{2})T < t < (n+1)T) \end{cases}$ | E1.1.12 |
| **2.86** | $\frac{a}{p^2+a^2}\coth\left(\frac{\pi p}{2a}\right).$ | $\lvert\sin at\rvert \quad (a > 0)$ | E1.5.8 |
| **2.87** | $\sum_{n=1}^{\infty} k_n \frac{\exp(-a_{n-1}p) - \exp(-a_n p)}{p}$ | $k_n(a_{n-1} < t < a_n) \quad (a_0 = 0)$ | E1.1.10 |
| **2.88** | $\frac{1-\mathrm{e}^{-p}}{p}\sum_{n=0}^{\infty} k_n \mathrm{e}^{-np}$ | $k_n \quad (n < t < n+1)$ | E5.2.3 |
| **2.89** | $\mathrm{e}^{-ap}$ | $\delta(t-a) \quad (a \geqslant 0)$ | E1.1.11 |

# List of Definitions

# Answers to Additional Examples

**E1.5.1** (i) $\dfrac{6}{p^4}+\dfrac{6}{p^3}+\dfrac{2}{p^2}$; (ii) $\dfrac{p+2}{(p+1)^2}$; (iii) $\dfrac{2p(p^2-3)}{(p^2+1)^3}$;

(iv) $\dfrac{6}{(p^2+1)(p^2+9)}$; (v) $\dfrac{80-4p^2}{(p^2-4)^3}$; (vi) $\dfrac{2p}{p^4+4}$.

**E1.5.2** (i) $\frac{1}{2}(1-e^{-t})^2$; (ii) $\frac{5}{2}e^{-3t}-\frac{3}{2}e^{-t}$;
(iii) $e^{-2t}(\cos t+\sin t)$; (iv) $2\cos t-1$;
(v) $2e^t-t-2$; (vi) $(2t-1)e^t+1$.

**E1.5.6** $\dfrac{1}{p^2}+\left(\dfrac{a}{p+1}-\dfrac{1}{p^2}-\dfrac{a}{p}\right)e^{-ap}$.

**E1.5.7** $\dfrac{a}{p(e^{ap}-1)}$.

**E1.5.8** $\bar{v}(p)=\dfrac{1+e^{-\pi p}}{p^2+1}$, $\quad\bar{u}(p)=\dfrac{\coth(\frac{1}{2}\pi p)}{p^2+1}$.

**E1.5.9** $\bar{v}(p)=\dfrac{1}{p^2}-\left(\dfrac{1}{p^2}+\dfrac{a}{p}\right)e^{-ap}$, $\quad\bar{u}(p)=\dfrac{1}{p^2}-\dfrac{a}{p(e^{ap}-1)}$.

**E1.5.23** $\alpha\tan^{-1}\left(\dfrac{\alpha}{p}\right)-\frac{1}{2}p\log\left(1+\dfrac{\alpha^2}{p^2}\right)$.

**E1.5.24** $p^{-1}\log(p+1)$, $\quad p^{-1}\cot^{-1}p$.

**E1.5.25** (i) $\frac{1}{4}e^t-(\frac{1}{2}t+\frac{1}{4})e^{-t}$; (ii) $\frac{1}{2}t\sinh t-\cosh t+1$;
(iii) $\frac{1}{2}(t-1)\sin t+\frac{1}{2}\cos t-\frac{1}{2}e^{-t}$;
(iv) $(3\cos 2t+\sin 2t)e^{-2t}-(3\cos 2t+4\sin 2t)e^{-4t}$.

**E1.5.27** $\dfrac{2\alpha}{t}(\sinh\alpha t+\sin\alpha t)-\dfrac{2}{t^2}(\cosh\alpha t-\cos\alpha t)$.

**E2.5.1** (i) $-\frac{14}{3}+4t-3t^2+e^{-2t}+4e^{-t}+\frac{2}{3}e^{3t}$;
(ii) $2\cos t+\sin t-3e^{-t}+e^{-3t}$;

(iii) $1-2e^{-t}+e^{t}\cos 2t$; (iv) $e^{2t}-(t+1)e^{t}$;

(v) $\frac{1}{4}(t-1)+\frac{1}{4}(2t^2+9t+9)e^{-2t}$;

(vi) $(\frac{7}{6}t-4)\cosh 3t-\frac{1}{6}\sinh 3t$; (vii) $\frac{1}{4}e^{t}-\frac{1}{4}(2t+1)e^{-t}$;

(viii) $(\frac{1}{2}t-1)e^{t}+\cos t+\frac{3}{2}\sin t$;

(ix) $\frac{1}{2}(t\cos t-\sin t)+\frac{1}{3}\sinh 3t$.

**E2.5.2** (i) $u(t)=\frac{1}{2}e^{t}-\frac{1}{10}e^{-3t}-\frac{2}{5}e^{2t}$, $v(t)=-\frac{1}{4}e^{t}-\frac{3}{20}e^{-3t}+\frac{2}{5}e^{2t}$;

(ii) $u(t)=t\cos t-\sin t$, $v(t)=2t\sin t+\cos t$;

(iii) $u(t)=e^{t}(\cos t-\sin t)$, $v(t)=e^{t}(\cos t+\sin t)$;

(iv) $u(t)=-\frac{7}{2}\cos(\sqrt{2}t)+2\sqrt{2}\sin(\sqrt{2}t)+6\cos(\sqrt{3}t)$
$-\frac{4}{3}\sqrt{3}\sin(\sqrt{3}t)-\frac{3}{2}\cos 2t$,

$v(t)=7\cos(\sqrt{2}t)-4\sqrt{2}\sin(\sqrt{2}t)$
$-9\cos(\sqrt{3}t)+2\sqrt{3}\sin(\sqrt{3}t)+2\cos 2t$;

(v) $u(t)=3-2(t+1)e^{-t}$, $v(t)=(t+1)e^{-t}$;

(vi) $u(t)=(\frac{1}{2}t\sin t+\cos t)e^{-t}$, $v(t)=-\frac{1}{2}t\sin t\, e^{-t}$

**E2.5.3** (i) $\frac{1}{2}(\sin t-t\cos t)$ $(0\leqslant t\leqslant\pi)$, $-\frac{1}{2}\pi\cos t$ $(t\geqslant\pi)$;

(ii) $\frac{1}{2}t-\frac{3}{4}+e^{-t}-\frac{1}{4}e^{-2t}$ $(0\leqslant t\leqslant a)$,
$[1+(at-a^2-1)e^{a}]e^{-t}+[-\frac{1}{4}+(\frac{1}{4}+\frac{1}{2}a)e^{2a}]e^{-2t}$
$(t\geqslant a)$.

**E2.5.4** (i) $(1-\frac{1}{3}t)e^{-t}$; (ii) $1-\dfrac{2}{\sqrt{3}}\sin\left(\dfrac{\sqrt{3}}{2}t+\dfrac{\pi}{3}\right)e^{-t/2}$;

(iii) $e^{-t}\cos t$; (iv) $1-\cos t-\tan\frac{1}{2}T\sin t$
$(T\neq(2n+1)\pi)$; (v) $\frac{1}{2}\sin t+(\frac{1}{4}\pi-\frac{1}{2}t)\cos t$.

**E2.5.6** $\frac{25}{21}e^{-4t}-\frac{4}{21}e^{-25t}$.

**E2.5.7** (i) $10^{-4}E_0[1-(\cos 60t+\frac{4}{3}\sin 60t)e^{-80t}]$;

(ii) $(E_0/16\,000)\{\sin(100t+\alpha)$
$-[\sin\alpha\cos 60t+(\frac{5}{3}\cos\alpha+\frac{4}{3}\sin\alpha)\sin 60t]e^{-80t}\}$.

**E2.5.8** $E_0(\frac{1}{8}-\frac{1}{12}e^{-2t}-\frac{1}{24}e^{-28t/5})$.

**E2.5.9** $\frac{7}{10}\sin 6t-\frac{2}{5}\cos 6t+\frac{1}{2}e^{-12t}-\frac{1}{10}e^{-18t}$.

**E2.5.10** $E_0(-\frac{144}{1885}\cos 8t-\frac{12}{1885}\sin 8t-\frac{3}{2405}e^{-t}+\frac{1083}{13949}e^{-38t})$.

**E2.5.11** $[14\cos(\sqrt{\frac{13}{12}}t)-\sqrt{\frac{12}{13}}\sin(\sqrt{\frac{13}{12}}t)]e^{-t/2}-4e^{-2t}$.

**E2.5.12** (i) $2t+e^{-t}$; (ii) $t\cos t$.

**E3.5.1** (i) $\frac{1}{2} - e^{-t} + \frac{1}{2}e^{-2t}$; (ii) $2e^t - t - 2$;
(iii) $\frac{1}{2}t^2 - 2t + 3 - (t+3)e^{-t}$;
(iv) $1 - \frac{1}{2}(t-1)\sin t - \frac{1}{2}(t+2)\cos t$.

**E3.5.3** (i) $\sum_{n=0}^{\infty} \frac{(-1)^n e^{-(\alpha+n)t}}{\Gamma(\beta-\alpha-n)n!} \quad \left[= \frac{e^{-\alpha t}}{\Gamma(\beta-\alpha)}(1-e^{-t})^{\beta-\alpha-1}\right]$;

(ii) $\sum_{n=0}^{\infty} (e^{-(\beta+n)t} - e^{-(\alpha+n)t}) \quad \left[= \frac{e^{-\beta t} - e^{-\alpha t}}{1-e^{-t}}\right]$;

(iii) $\sum_{n=0}^{\infty} (-1)^n (2n+1)\pi \exp[-(n+\frac{1}{2})^2\pi^2 t]$;

(iv) $1 - \sum_{n=0}^{\infty} (4n^2\pi^2 t - 2)\exp(-n^2\pi^2 t)$.

**E3.5.4**

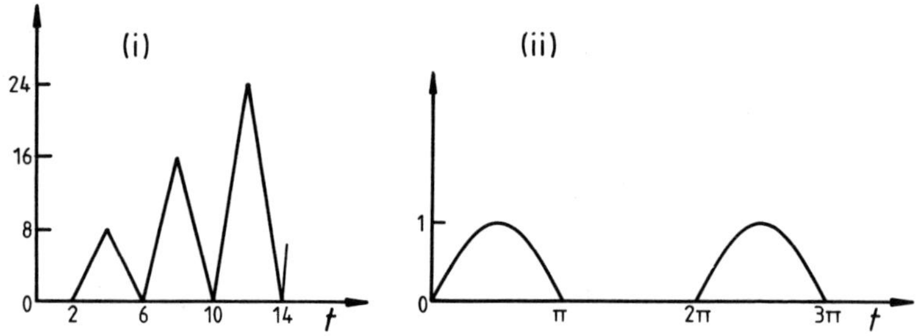


**E3.5.5** (ii) $4\pi^{-1/2}t^{-3/2} \sum_{n=1}^{\infty} n^2 \exp(-n^2/t)$.

**E3.5.9** $u(t) \sim \frac{2}{3}e^t - \sum_{n=0}^{\infty} \frac{(-1)^n\Gamma(3n+\frac{5}{2})}{\pi t^{3n+5/2}}, \quad u(t) = \sum_{n=0}^{\infty} \frac{t^{(3n+1)/2}}{\Gamma(\frac{3}{2}n+\frac{3}{2})}$.

**E3.5.10** $u(t) \sim \frac{1}{2}\pi \cos t - \sum_{n=0}^{\infty} \frac{(-1)^n(2n)!}{t^{2n+1}}$,

$$u(t) = \sum_{n=0}^{\infty} \frac{(-1)^n t^{2n+1}}{(2n+1)!}[\psi(2n+2) - \log t].$$

**E3.5.11** $2\sqrt{\left(\frac{t}{\pi}\right)}\left(1 - \sum_{n=1}^{\infty} \frac{B_n(1-2^{-2n})}{n!(2n-1)t^{2n}}\right)$.

**E3.5.14** (i) $u(t) = \sum_{n=0}^{\infty} \frac{t^{n/2}}{n!\,\Gamma(\frac{1}{2}n+1)}$;

(ii) $\sum_{n=0}^{\infty} \frac{(-1)^n t^{n/2}}{n!\,\Gamma(\frac{1}{2}n+1)}$,

$$\frac{2}{\sqrt{(3\pi)}}(\tfrac{1}{4}t)^{-1/6}\exp[-\tfrac{3}{2}(\tfrac{1}{4}t)^{1/3}]\cos\left[\tfrac{3}{2}\sqrt{3}(\tfrac{1}{4}t)^{1/3}-\frac{\pi}{3}\right].$$

**E4.5.1** (i) $\frac{1}{2}xt^2 - \frac{1}{6}t^3 + \sin(x-t) \quad (t<x)$,
$\frac{1}{2}x^2t - \frac{1}{6}x^3\,(t>x)$;

(ii) $1 - e^{-t} \quad (t < \log(x+1))$,
$t - \log(x+1) + xe^{-t}(t > \log(x+1))$;

(iii) $t+1 \quad (t<\frac{1}{2}x^2), \quad \frac{1}{2}x^2 + \exp(t-\frac{1}{2}x^2) \quad (t>\frac{1}{2}x^2)$;

(iv) $xt + (\sinh t - t)e^{-x} \quad (t<x)$,
$xt + t - x + 1 - te^{-x} - e^{-t}\cosh x \quad (t>x)$;

(v) $1 + x(t-1+e^{-t}) \quad (t<x)$,
$x(t-1+e^{-t}) + \cos(t-x) \quad (t>x)$;

(vi) $0 \quad (t<x), \quad t-x \quad (x<t<2x), \quad x \quad (t>2x)$;

(vii) $x + 2\sqrt{t}\,\mathrm{i\,erfc}\left(\frac{x}{2\sqrt{t}}\right)$;

(viii) $t - 4t\,\mathrm{i}^2\,\mathrm{erfc}\left(\frac{x}{2\sqrt{t}}\right)$.

**E4.5.2** $t - \frac{a}{r}\sum_{n=0}^{\infty}[(t-a_n)\mathrm{H}(t-a_n) - (t-b_n)\mathrm{H}(t-b_n)]$

$$= \begin{cases} t - 2m\dfrac{a}{c} & (b_{m-1} < t < a_m) \\ \left(\dfrac{a}{r}-1\right)\left((2m+1)\dfrac{a}{c} - t\right) & (a_m < t < b_m), \end{cases}$$

where

$$a_n = \frac{(2n+1)a - r}{c}, \qquad b_n = \frac{(2n+1)a + r}{c};$$

$$\frac{2a^2}{\pi^2 cr}\sum_{n=1}^{\infty}\frac{(-1)^{n-1}}{n^2}\sin\left(\frac{n\pi r}{a}\right)\sin\left(\frac{n\pi ct}{a}\right).$$

**E4.5.3** $\sum_{n=0}^{\infty}\{\sin[\omega(t-a_n)]\,\mathrm{H}(t-a_n) - \sin[\omega(t-b_n)]\,\mathrm{H}(t-b_n)\}$,
where $a_n = \dfrac{2na+x}{c}$, $b_n = \dfrac{(2n+2)a-x}{c}$;

$$\frac{\sin[\omega(a-x)/c]}{\sin(\omega a/c)}\sin\omega t + 2\frac{\omega a}{c}\sum_{n=1}^{\infty}\frac{\sin(n\pi x/a)\sin(n\pi ct/a)}{(\omega a/c)^2 - n^2\pi^2}$$

($\omega a/\pi c$ non-integral),

$$\frac{2k}{\pi}\sum_{\substack{n=1\\ n\neq k}}^{\infty}\frac{\sin(n\pi x/a)\sin(n\pi ct/a)}{k^2-n^2}+\left[\left(\frac{1}{2k\pi}-\frac{ct}{a}\right)\sin\left(\frac{k\pi x}{a}\right)\right.$$

$$\left.+\left(1-\frac{x}{a}\right)\cos\left(\frac{k\pi x}{a}\right)\right]\sin\left(\frac{k\pi ct}{a}\right)\qquad(\omega a/c=k\pi).$$

**E4.5.6** $$\mathrm{e}^{t-x}+\operatorname{erf}\left(\frac{x}{2\sqrt{t}}\right)-\tfrac{1}{2}\mathrm{e}^{t+x}\operatorname{erfc}\left(\frac{x}{2\sqrt{t}}+\sqrt{t}\right)$$
$$-\tfrac{1}{2}\mathrm{e}^{t-x}\operatorname{erfc}\left(\frac{x}{2\sqrt{t}}-\sqrt{t}\right).$$

**E4.5.8** (i) $\mathrm{I}_0(2\sqrt{(xy)})-\int_0^x \mathrm{I}_0(2\sqrt{(wy)})\mathrm{e}^{w-x}\,\mathrm{d}w$;

(ii) $\mathrm{e}^{-x-y}+\int_0^y \mathrm{I}_0(2\sqrt{(xz)})\mathrm{e}^{z-y}\,\mathrm{d}z$.

**E4.5.11** $-8h/\alpha_n^3\,\mathrm{J}_0'(\alpha_n)$.

**E4.5.12** $$\tfrac{1}{2}U\left[\operatorname{erfc}\left(\frac{y+Vt}{2\sqrt{(\nu t)}}\right)+\exp\left(-\frac{Vy}{\nu}\right)\operatorname{erfc}\left(\frac{y-Vt}{2\sqrt{(\nu t)}}\right)\right].$$

**E4.5.13** $$\frac{a^2\omega}{r}\operatorname{erfc}\left(\frac{r-a}{2\sqrt{(\nu t)}}\right);\quad \frac{a^3\omega}{r^2}\operatorname{erfc}\left(\frac{r-a}{2\sqrt{(\nu t)}}\right);$$
$$\frac{a^5\omega}{r^4}\left[\operatorname{erfc}\left(\frac{r-a}{2\sqrt{(\nu t)}}\right)+\frac{r-a}{a}\exp\left(\frac{r-a}{a^2}+\frac{\nu t}{a^2}\right)\right.$$
$$\left.\times\operatorname{erfc}\left(\frac{r-a}{2\sqrt{(\nu t)}}+\frac{\sqrt{(\nu t)}}{a}\right)\right].$$

**E5.6.1** (i) $t+2+2(t-1)\mathrm{e}^t$;

(ii) $g(t)-2\int_0^t g(s)\mathrm{e}^{t-s}[\cos(t-s)-\sin(t-s)+\tfrac{1}{2}]\,\mathrm{d}s$;

(iii) $1+\tfrac{1}{2}t^2$; (iv) $\pi^{-1/2}\mathrm{e}^{-t/2}\mathrm{I}_0(\tfrac{1}{2}t)$.

**E5.6.2** (i) $(5-4\cos t-2\sin t)\mathrm{e}^{-t}$; (ii) $\mathrm{J}_1(2t)/t$.

**E5.6.3** $$\tfrac{2}{3}\mathrm{e}^t+\sum_{n=0}^{\infty}\frac{(-1)^n\Gamma(3n+\tfrac{3}{2})}{\pi t^{3n+3/2}}.$$

**E5.6.5** (i) $\frac{1}{3}\sum_{m=2}^{\infty}[1-(-2)^{m-1}](t-m)\mathrm{H}(t-m)$;

(ii) $\sum_{m=1}^{\infty}[1+(m-1)\mathrm{e}^{-(t-m)}]\mathrm{H}(t-m)$.

**E5.6.6** (i) $\frac{1}{2}[(1+\sqrt{2})^n+(1-\sqrt{2})^n]$;

(ii) $a_{2m}=-2m, \quad a_{2m+1}=3m+1$.